Springer Tracts in Natural Philosophy

Volume 14

Edited by B.D. Coleman

Co-Editors: R. Aris · L. Collatz · J.L. Ericksen · P. Germain

M.E. Gurtin · M. M. Schiffer · E. Sternberg · C. Truesdell

James F. Bell

The Physics of Large Deformation of Crystalline Solids

With 166 Figures

Springer-Verlag New York Inc. 1968

James F. Bell

Professor of Solid Mechanics
The Johns Hopkins University
Baltimore, USA

ISBN 978-3-642-88442-9   ISBN 978-3-642-88440-5 (eBook)
DOI 10.1007/978-3-642-88440-5

Title No. 6742

*This monograph is dedicated to my wife, Perra,
who has given life its meaning; and to our daughter, Jane,
and our son, Christopher*

Preface

Historically, a major problem for the study of the large deformation of crystalline solids has been the apparent lack of unity in experimentally determined stress-strain functions. The writer's discovery in 1949 of the unexpectedly high velocity of incremental loading waves in pre-stressed large deformation fields emphasized to him the pressing need for the independent, systematic experimental study of the subject, to provide a firm foundation upon which physically plausible theories for the finite deformation of crystalline solids could be constructed.

Such a study undertaken by the writer at that time and continued uninterruptedly to the present, led in 1956 to the development of the diffraction grating experiment which permitted, for the first time, the optically accurate determination of the strain-time detail of non-linear finite amplitude wave fronts propagating into crystalline solids whose prior history was precisely known. These experimental diffraction grating studies during the past decade have led to the discovery that the uniaxial stress-strain functions of 27 crystalline solids are unified in a single, generalized stress-strain function which is described, much of it hitherto unpublished, in the present monograph.

The detailed study of over 2,000 polycrystal and single crystal uniaxial stress experiments in 27 crystalline solids, in terms of the variation of a large number of pertinent parameters, has provided new unified patterns of understanding which, it is hoped, will be of interest and value to theorists and experimentalists alike.

The writer would like to acknowledge the assistance at different times and in varying degrees, of numerous research assistants and graduate students. He would like particularly to thank WILLIAM SOPER and SUSUMU ISHIWATA for their assistance during the frustrating early years of this experimental development; and JOHN SUCKLING and JOHN GOTTSCHALK, successive later research assistants, for their able and consistently competent cooperation in a laborious and often repetitive experimental program. Gratitude is expressed to Mrs. PATRICIA JONES, Mrs. VIRGINIA FRANZEN FISHER, and Mrs. AMANDA HOLLRAH for their enthusiasm and painstaking care in carrying out the major portion of over 2,000 individual test calculations; to Mrs. FAITH PAQUET, artist, and Mrs. MARY COREY, calculator, for their sustained concern that all of the drawings and the compiled calculations in this monograph should be accurate; to Mrs. HILLARY HOLDEN, Mrs. LESLIE ROBERTS, and Mrs. HELEN RUTHERFORD who spent some of their time in America accurately contributing to the voluminous laboratory calculations; and to Mrs. ANNE FRANCIS for her efficient and stoical typing and retyping of the manuscript.

The author would like to record his gratitude to his wife, PERRA, who, while pursuing her own scholarly activities, always has found time to contribute her editorial perspective and other invaluable labors to the author's manuscripts over many, many years.

The author is very appreciative of Mr. BERNARD BAKER, machinist, and his group, for their careful preparation, to exacting specification of over 2,000 specimens.

Of the many former graduate students, some of whose data is noted in this monograph, the author would like especially to thank Drs. GORDON L. FILBEY, WILLIAM GILLICH, and WILLIAM HARTMAN, who have read and helpfully criticized the manuscript.

Appreciation is expressed by the author for the years of listening and scholarly criticism of his dear colleagues: Professors EDWIN R. FITZGERALD, ROBERT POND, JOHN STRONG, CLIFFORD TRUESDELL, and, particularly, Professor JERALD L. ERICKSEN, who not only has been a friend and most valuable critic for many years, but also has read and criticized this manuscript in detail.

Much of the research upon which this monograph is based was carried out over the past many years under the sponsorship of the United States Air Force, Office of Scientific Research. Many of the wave propagation studies of Chapter II were sponsored by the United States Army, Army Research Office (Durham). The very early diffraction grating studies and much of the more recent high velocity research related to the transition structure has been sponsored by the United States Army, Ballistics Research Laboratories, Aberdeen Proving Ground.

And, finally, the author wishes to acknowledge and thank Drs. J. D. B. KING, A. GENECIN, and W. B. KOUWENHOVEN, whose medical wisdom provided the second life in which this monograph was written.

Baltimore, November 1967   JAMES F. BELL

Contents

Contents

Chapter I

Introduction

On a cylindrical solid the simultaneous measurement of an axial force and the axial deformation it produces is conceptually one of the simplest experiments in experimental physics. That a vast literature has been written during the last two centuries to describe the functional relation between load and deformation in such experiments attests to the complexity of nature. In this monograph, large deformation uniaxial stress experiments are seen as providing the opportunity for studying the finite distortion of crystalline solids under conditions which minimize the influence of compressibility upon the deformation.

Non-linear continuum theories of solids are seldom of interest when constrained by the lack of generality implied in considering only a single stress state, such as uniaxial stress. The related but separate field of the non-linear continuum experimentalist, on the other hand, finds generality in just such experimentally simple situations since they permit a controllable cross comparison of a variety of solids and deformation histories. The high pressure research of P. W. BRIDGMAN in the single stress state of hydrostaticity offers a classical example. The uniaxial stress state for finite deformation in cylindrical isotropic polycrystalline solids phenomenologically represents a nearly pure distortion. One obtains a predominantly equivoluminal distortion whether the experiment is a quasi-static measurement or a one-dimensional non-linear wave front measurement. For the uniaxial stress state, unlike the single stress state hydrostaticity situation, finite amplitude wave fronts and quasi-static uniform strain occur in the same form.

Hydrostatic experiments and uniaxial stress experiments may be considered to represent two large deformation extremes contrasting a situation of high stress and small strain with a situation of high strain and small stress. For the former, ideally one obtains a finite change of volume without distortion and for the latter, a finite distortion without a change of volume. This comparison of conditions is shown in Fig. 1.1 for polycrystalline aluminum. As will be shown below, it is interesting that so many similarities have been found when comparing the observed finite deformation behavior of these two extremes.

The finite distortion of crystalline structures can be accomplished only by the major realignment of atomic components. The scale of such adjustment may vary from lattice dimensions to macroscopic mosaics, with more than one mechanism in operation for different degrees of deformation. Despite the complexity revealed by large deformation experiments in the form of multiple deformation modes, second order transitions, Portevin - le Chatelier instabilities, multiple-elastic moduli and the like, the study of the mechanics of distortion is essential to an understanding of the structure of solids.

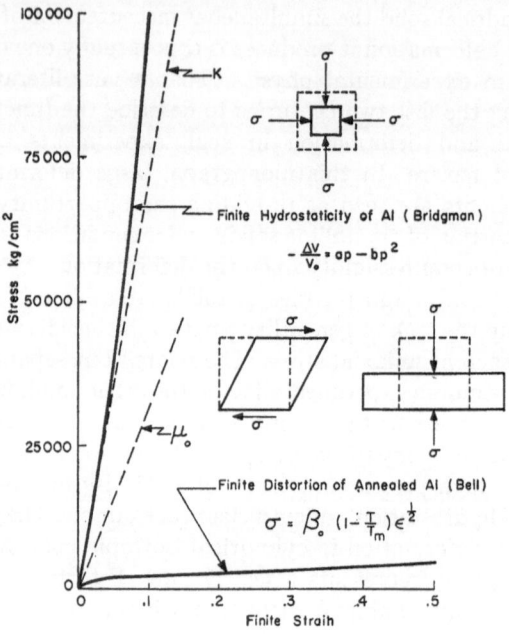

Fig. 1.1. A comparison of finite hydrostaticity and finite distortion stress-strain functions for polycrystalline aluminum

The major experimental difficulty of a generalized single stress state study is to obtain and maintain the prescribed condition for each of the various experiments being considered. A few examples of such experimental difficulties are: the maintenance of isotropy in quasi-static hydrostaticity experiments; the large distortion which always accompanies plane shock dilatation waves in dynamic hydrostaticity experiments; grip effects in quasi-static uniaxial stress experiments; or the obtaining of one dimensionality for finite amplitude distortion wave front experiments.

Uniaxial stress experiments described in this monograph have been performed over the entire temperature scale from within 4° K of absolute zero to within 20° K of the melting point. Axial loads applied by compression impact in time intervals much less than a microsecond have produced strains in the immediate vicinity of the point of application of $7^0/_0$ or $8^0/_0$ in a single microsecond. (In the present study, the highest experimentally measured strain rate in such situations is $\dot{\varepsilon} = 7 \times 10^4 \sec^{-1}$.) One may contrast the strains produced by such rapid loading with the production of deformations of a fraction of a percent over a period of many hours, or even years, under the influence of a constant quasi-static force. (The lowest strain rate considered experimentally in the present study is $\dot{\varepsilon} = 10^{-9} \sec^{-1}$.) Between these limits of an instantly applied constant force maintained for a few microseconds and a maintained quasi-static constant force, an infinitude of force-time histories is possible.

By far the most common uniaxial stress quasi-static experiment is that in which the deformation history is prescribed. The usual choice is to require that the uniform axial strain shall increase linearly with time. The experimental problem is to determine the axial load-time history necessary to produce this prescribed deformation-time relationship. With respect to distortional deformation studies, a more interesting uniaxial stress quasi-static experiment is that in which the axial load history is prescribed either as a linear function of time or as a series of intermittent incremental loadings. In such a situation, which more closely resembles that of one-dimensional finite amplitude wave propagation, the experimental unknown is the axial strain-time history. The magnitude of the constant stress rate or constant strain rate may be arbitrarily chosen until a strain rate of approximately 1 $\sec^{-1}$ is reached when inertia effects begin to predominate. Above this strain rate of 1 $\sec^{-1}$ the deformation is controlled by the dynamic wave propagation characteristics of the solid and will vary with position and time, as dictated by the nonlinear mechanics. Load and deformation histories no longer are subject to arbitrary specification as in quasi-static experiments.

Considered in the present monograph are nearly 2,000 individual uniaxial stress experiments in 19 crystalline elements and several binary combinations. The experiments include many different force-time histories, ambient temperatures, crystalline purities, prior deformation histories, and metallurgical states. These metallurgical variables include differences in methods of specimen preparation for otherwise similar crystalline solids. From an examination of all of these data a generalized large deformation behavior has been discovered; it is presented here for the first time in detailed form. The generalized behavior is characterized by a finite deformation mode and transition structure in terms of which one may express the large distortional deformation of all of the elements and binary combinations considered. The major problem of the finite distortion of crystalline solids is seen to be in the hitherto unrecognized stability properties of a large distortional deformation mode and transition structure rather than in viscous properties.

1*

The polycrystalline solids whose finite distortion is considered in the present study are Al, Au, Ag, Pb, Ni, Ir, Rh, Cu, Fe, Cr, Mo, Ta, Nb, Mg, Zn, Re, Y, Ge, U; the binary combinations Cu-30 Zn, Mo-50 Re, and a number of low percentage alloys of the above group. The single crystal, resolved shear deformation of the cubic crystals Al, Ag, Au, Ni, Pb, Cu, Fe, Ta, Mo, Na-Cl, and various percentage binary combinations of Ag-Au and Ni-Co also have been studied.

In every instance the stress-strain function describing the large deformation behavior is found to be parabolic. The choice of stress definition and strain measure is of course arbitrary. It is important to note at the outset, however, that the experimentally observed parabolicity of the polycrystal occurs when the uniaxial stress-strain function is expressed in *nominal* form, referring to the undeformed state of the material.

All parabola coefficients have been shown to be linearly dependent on the temperature and to be proportional to the zero-point isotropic elastic modulus $\mu(0)$ multipled by a dimensionless universal constant $B_0 = 0.0280$. The deformation mode is, in every instance, designated by a discrete factor $(2/3)^{r/2}$ where $r = 0, 1, 2, 3, ---$.

Thus, irrespective of strain rate, purity, temperature, grain size, or previous metallurgical history, all 19 polycrystalline elements and both polycrystalline binary combinations have an uniaxial stress-strain function given by Eq. (1.1)

$$\sigma = (2/3)^{r/2}\mu(0)\, B_0(1 - T/T_m)(\varepsilon - \varepsilon_b)^{1/2} \qquad (1.1)$$

where σ is the nominal uniaxial stress, ε the nominal uniaxial strain, and ε_b is the predictable parabola intercept on the strain abscissa. One notes that T_m, $\mu(0)$, B_0, T, and ε_b are all known in Eq. (1.1); the only unspecified parameter is the discrete integral value of the deformation mode index r. Through the stress and strain ratios of the TAYLOR (1938) and BISHOP and HILL (1951) aggregate theory, the stage III resolved deformation of all of the cubic single crystals listed above will be shown to have shear stress and strain functions expressible in similar form.

What is of perhaps greatest significance in this large deformation generalization whose existence will be established in the remainder of this monograph, is the discovery of second order transitions in the form of slope discontinuities designated by discrete changes of the integer r. These transitions are shown to be accompanied by a similar discrete shift in the value of the zero-point isotropic modulus $\mu(0)$. The stable zero-point isotropic elastic shear moduli of 58 elements are found to occur in the form

$$\mu(0) = (2/3)^{s/2}(2/3)^{p/4}A \qquad (1.2)$$

where $s = 1, 2, 3, ---$; p is a constant structure factor, $p = 0$ or 1;

and A is a universal constant, $A = 2.89 \times 10^4$ kg/mm². As will be shown below, the integer r designates the discrete multiple elastic moduli described here for the first time.

The value of the mode index $r = 1$ was arbitrarily assigned to the resolved stage III parabola coefficients of Al, Pb, and Cu at 4.2° K. These were the lowest experimental values of r found for each metal, as will be shown in Chapter IV below. The stage III data of NOGGLE and KOEHLER (1957) at 4.2° K in aluminum provide a parabola coefficient of 13.38 kg/mm²; the stage III parabola coeffitient average of five experiments of BOLLING, HAYS, and WIEDERSICH (1962) in lead at 4.2° K is 3.22 kg/mm²; and the stage III parabola coefficient of four tests of MITCHELL and THORNTON (1963) in Cu at 4.2° K is 22.14 kg/mm². Each of these averaged stage III parabola coefficients is in the ratio of zero-point isotropic shear moduli for Al, Pb, and Cu, namely 3,110 kg/mm², 750 kg/mm², and 5,080 kg/mm² respectively.

Assigning a common value of mode index $r = 1$ to these 4.2° K data sets the large deformation mode index pattern for all of the other single crystal and polycrystal stress-strain parabola coefficients described in this monograph. The ratio of parabola slopes at a transition is specified by a shift in the integral value of r which, in turn, designates instantaneous isotropic shear moduli values for a given element. The integer s designates the discrete distribution of zero-point isotropic shear moduli among the different elements.

It is the purpose of this monograph to develop the experimental details of this new large deformation generalization and its mode and transition structure. Unless otherwise designated, the experiments have been performed in the writer's laboratory. Beginning with test No. 1 in September 1958, nearly all of the experiments of the writer and his students have been consecutively numbered. At this writing, nine years later, the number of experiments in this series exceeds 1,400.

This systematic study of the finite distortion of crystals by the writer has been in progress for eighteen years, during which time numerous papers have described many preliminary aspects of the present generalization (BELL, 1961b; 1962a, b; 1963a, 1964; 1965a, b; 1967a; BELL and WERNER, 1962). Most of the research described below, which has grown out of this earlier research, is being presented for the first time in the present monograph.

Plan of Monograph

Experimental research performed to reveal new patterns of understanding in mathematically oriented fields such as non-linear continuum mechanics or crystal physics is bounded by two limiting logics. On the one hand unrestrained empiricism with its infinitude of curve fitting alternatives and inscrutable models does not provide the means for distinguishing between the trivial and the significant. Theory-dominated experiment, on the other hand, which is performed primarily to distinguish between plausible hypotheses, is limited by the logic of mathematical tractability. Experimental discovery in this second situation is constrained within the imaginative limits of the theorist who all too often is insensitive to, and not particularly interested in, experiment. In view of the fact that sooner or later experimental data are usually produced to "verify" most plausible analytically successful hypotheses, independent experimental study is essential to retain contact with the physical problem. In this monograph, an effort has been made to present the experimental data in the first nine chapters with a minimum of speculation and interpretation. Such matters are considered in the final chapter.

At some point between the correlation of a few pieces of data in an empirical framework and the demonstration of the interrelation of hundreds of different measurements in terms of the variation of numerous fundamental quantities, an experimental generalization is established. Such a development may ultimately be more conveniently expressed in terms other than those in which it was first discovered or its significance may be variously interpreted at different times, but if the experiments are sound, and the measurements adequate, the phenomena must be considered as part of the recognized behavior of the field and must be retained in any satisfactory theoretical explanation. Chapters II, III, IV, and VIII describe the uniaxial finite amplitude wave experiments and the uniaxial quasi-static experiments of 19 polycrystalline elements and several binary combinations in terms of the generalization of Eq. (1.1) which evolved from the consideration of these same data. Chapter V and Chapter VI describe the discrete distribution of zero-point elastic moduli, both among the elements, Eq. (1.2), and as multiple values for individual elements. The proportionality between large deformation parabola coefficients and these discretely distributed elastic moduli is further described in Chapter VII on the PORTEVIN - LE CHA-TELIER (1923) instabilities characterizing slow rate deformation, and in Chapter IX which considers the high strain rate transition velocities and slow strain rate slope discontinuities in the stress-strain functions of Chapters II, III, IV, and VIII. The various pieces are assembled and analyzed in Chapter X.

Chapter II

Finite Amplitude Wave Propagation Experiments

The finite distortional deformation generalization described in this monograph was first discovered during the analysis of observations from finite amplitude wave propagation experiments. The present description of uniaxial stress distortional deformation begins with the discussion of these data.

The distinguishing feature of finite amplitude distortion wave studies is the development of finite strain in microsecond time intervals. In a time scale from one microsecond to a few hundred microseconds, complicated slow time phenomena such as grain boundary slippage, long time creep, etc. are suppressed with a resulting major simplification of the large deformation behavior. In these relatively simplified terms the general underlying unity of the stress-strain function governing a variety of elements and binary combinations became apparent. In this short time scale simplification also was achieved for the study of the effects upon large deformation of purity, grain size, viscosity, ambient temperature, differences in melting point, and magnitude of the initial deformation.

When the deformation occurs at loading rates sufficiently high to introduce inertia effects, the prescription of an arbitrary loading history or deformation history at each point in the solid is no longer possible. To obtain stress-strain functions for large distortional deformation during finite amplitude wave propagation, it has been necessary to adopt a logical approach which does not require that the answer be guessed in advance and in which it is possible from the experiment itself to ascertain, without question, that the necessary uniaxial stress conditions prevail. Such an experiment is the measurement of finite dynamic strain during the symmetrical free-flight axial impact of identical cylinders.

In this situation, when the impact velocity is sufficiently high, identical finite amplitude waves propagate in opposite directions from the common interface. This symmetry not only insures a constant velocity impact but also provides assurance that there shall be no interchange of electrical, mechanical, or thermal energy between specimens. Initial conditions may thus be prescribed with confidence, without the necessity of recourse to auxiliary empirical arguments. The radial expansion of both specimens is the same at the contact surface;

hence, one need have no special concern with respect to interface friction. Measurements of strain-time and particle velocity-time histories at numerous positions along long cylinders in such a symmetrical free-flight impact experiment allow a description of the space and time distribution of the highly dispersive finite amplitude wave front which has been produced.

These measurements provide finite strain and particle velocity wave speeds for all amplitudes at any position along the specimen. Without reference to any non-linear wave theory or the *a priori* assumption of any constitutive relation, one may note whether or not wave speeds for a given amplitude of strain or for a given amplitude of particle velocity are constant as the wave front propagates down the rod. One also may note, observationally, the relation between finite strain and particle velocity at any position along the rod, and in particular, in terms of the measured slope of the finite strain-time curve; i.e., the strain rate $\dot{\varepsilon}$. It is thus possible to construct, on purely empirical grounds, a detailed description of the functional relationship between strain ε, normalized particle velocity $\dot{u}$, and strain rate $\dot{\varepsilon}$ for the non-linear wave front by choosing an appropriate distance from the impact face and strain amplitude for any strain rate of interest.

The experimental program from which these observations were made, had been made possible in 1956 by the development of the writer's diffraction grating technique for the determination of finite strain at very high strain rates over the necessary extremely short gauge lengths (BELL, 1956b, 1958, 1960a, 1962b, 1967a).

Measuring dynamic finite strain by means of this new experimental method, the writer found that in every one of a large number of metals, wave speeds at each strain amplitude were constant with the value of the constant differing from one amplitude to another. This behavior was observed only after the non-linear wave front had propagated at least one diameter from the impact face (BELL, 1960b, 1961b). From a comparison of particle velocity measurements obtained in an optical displacement experiment, also developed for large deformation wave studies by the writer at approximately the same time as the diffraction grating technique, it was found that for each value of strain amplitude there was invariably associated a specified value of particle velocity; i.e., the particle velocity data also provided different constant wave speeds for each amplitude beyond the first diameter (BELL, 1961c). In stating this invariable relationship between particle velocity and strain in a particular experimental situation, reference, of course, must be made to the original velocity of the specimen prior to impact.

Further confirmation of the existence of an invariable relationship between finite strain and particle velocity was obtained through the comparison of maximum strains and maximum particle velocities (BELL, 1960b, 1961b). The maximum particle velocities were, of course, given as one-half the hitter velocity in the symmetrical free-flight impact situation. The impact velocity could be changed by varying the hitter velocity. Beyond the first diameter, for a prescribed maximum particle velocity, the maximum strain at all positions along the rod was identical. Even if there were no non-linear wave propagation theories available, such a finite amplitude wave propagation behavior as that just described demonstrates experimentally that this phenomenon is independent of strain rate or viscosity.

The measured strain rates in these earlier finite amplitude wave experiments of the writer varied from $\dot{\varepsilon} = 70{,}000$ sec^{-1} to values of strain rate of the order of $\dot{\varepsilon} = 1$ sec^{-1}. Inertial effects began to be important at strain rates in excess of this latter value. Such experimental observations of finite strain-time and particle velocity - time histories, of course, do not by themselves provide the desired stress-strain function governing the large distortional behavior at these high strain rates. In 1942 G. I. TAYLOR and T. VON KARMAN, and in 1945 K. A. RAKHMATULIN, independently modified the 1859—1860 EARNSHAW and RIEMANN theory for progressive waves in a compressible gas to apply to the one dimensional uniaxial stress non-linear wave propagation in cylindrical specimens. Assuming stress to be a single value function of strain $\sigma(\varepsilon)$ for a one dimensional uniaxial stress wave, Eq. (2.1) becomes Eq. (2.2)

$$\varrho_0 \frac{\partial^2 u}{\partial t^2} = \frac{\partial \sigma}{\partial x} \tag{2.1}$$

$$\varrho_0 \frac{\partial^2 u}{\partial t^2} = \frac{d\sigma}{d\varepsilon} \frac{\partial^2 u}{\partial x^2} \tag{2.2}$$

where $\varepsilon = \dfrac{\partial u}{\partial x}$ and ϱ_0 is the mass density. This finite amplitude wave theory is applicable to non-linear wave propagation in a particular solid if the wave speeds $C_p(\varepsilon)$ are constant for each given strain amplitude and have the form prescribed in Eq. (2.3)

$$C_p(\varepsilon) = \sqrt{\frac{d\sigma}{d\varepsilon}\bigg/\varrho_0} = \frac{dx}{dt} = \text{Constant} \tag{2.3}$$

and if an invariable relation between particle velocity and finite strain in terms of these constant wave speeds is given in the form of the integral of Eq. (2.4), referring to an initial zero specimen velocity.

$$\dot{u} = \int_0^\varepsilon C_p(\varepsilon)\, d\varepsilon. \tag{2.4}$$

Eqs. (2.3) and (2.4) describe the experimentally observed behavior in each of the many crystalline solids studied. For each of these crystalline solids at all of the many temperatures and impact velocities for which finite strain-time and particle velocity - time histories have been determined, the required conditions of Eqs. (2.3) and (2.4) are met and it may be stated, therefore, that in each instance this finite amplitude wave theory is applicable. It should again be stated that these experimental conditions are in general only obtained *after* the non-linear wave front has propagated one specimen diameter from the impact face.

Having shown that this one-dimensional uniaxial stress wave propagation theory is applicable, one may, without additional assumption, determine the governing stress-strain function by integrating Eq. (2.3) thus obtaining Eq. (2.5)

$$\sigma = \int_0^\varepsilon \varrho_0 C_p{}^2(\varepsilon)\, d\varepsilon. \tag{2.5}$$

Using the procedure just described, in over a thousand impact experiments performed on a variety of crystal solids, the writer has studied the governing stress-strain function Eq. (2.5) in terms of crystal structure, ambient temperature, impact velocity, strain rate, specimen purity, polycrystalline grain size, etc. (BELL, 1960b, 1961b, 1962b, 1963a, 1964, 1965a, 1967a). A consideration of a possible relationship between the resulting dynamic stress-strain function at high strain rates and quasi-static stress-strain functions at low strain rates was a natural consequence.

These studies have included a comparison of the dynamic polycrystalline stress-strain function with resolved shear stress, resolved shear strain functions of cubic single crystals which in turn have led to the establishment of a close interrelation of the two in terms of the stress and strain ratios of the TAYLOR (1938) and BISHOP and HILL (1951) aggregate theory. This correlation between polycrystal and single crystal, which is described in considerable detail in later sections of the present monograph, has provided the experimental foundation for the generalized finite distortional deformation behavior referred to in the introduction.

To perform the dynamic experiments it is necessary to have an accurate method for determining large strain in microsecond time intervals over very short gauge lengths, with very fast rise times. The diffraction grating technique by means of which this is accomplished has been described in considerable detail in several earlier papers (BELL, 1956b, 1958, 1960a, 1962b, 1967a). The basic principle is very simple. The angles of the various diffraction images, produced when monochromatic light falls on a series of equally-spaced grooves ruled on a specimen surface, depend upon the spacing of the grooves. When the specimen is subjected to either tension or compression strain of any magnitude, either elastic or plastic, at any temperature up to the melting point, the spacing of the grooves is changed, resulting in a calculable change in the angle of diffraction for each order. Whether the strain be quasi-static or dynamic, the measurement of the angular changes of any two diffraction images will provide direct optically accurate measurement of both strain and surface angle.

Features which make the diffraction grating technique particularly well suited to the study of dispersive finite amplitude waves may be listed: (a) gauge lengths are from 0.001 in. to 0.005 in., depending upon the grain size of the polycrystal under investigation; (b) measurements have been made for strain from 20×10^{-6} to $170,000 \times 10^{-6}$ on the same gauge, which is integral with the material; (c) all measurements are made without any necessity for any electrical contact with the specimen; (d) rise times of 10^{-7} sec are being used for high-strain-rate studies; (e) accurate measurement of strain may be made over the entire temperature scale, up to the melting point.

In dynamic plastic wave propagation studies, the extremely short gauge length is of great importance. Plastic waves are highly dispersive and, without a detailed *a priori* knowledge of the strain-time history, it is not possible to integrate over large gauge lengths. Thus, it has been possible for the present writer to make measurements of dynamic plastic strain with an 0.001 in. long diffraction grating located 0.020 in. from the impact face where, even at relatively low impact velocities, strain rates as high as 25,000 sec^{-1} have been observed experimentally. Such a measurement in annealed Al is shown in Fig.2.1 at 0.05 cm from the impact face for an impact velocity of only 2,650 cm/sec.

The observation of dynamic strain requires that changing angles of diffraction be observed in μ-sec time intervals. This has been accomplished by allowing the images to fall upon V-shaped slits in front of 5 in. photomultiplier tubes whose electrical output depends upon the total amount of light observed. Because the diffraction grating undergoes a displacement during the strain, it is necessary to have light fields which are larger than the grating. Because some diffuse light from the polished surfaces adjacent to the grating enters the photomultiplier tubes, it is necessary to calibrate the grating prior to use in order that the photomultiplier output may be correlated with the change in diffraction angle. The present writer has developed several techniques for making this calibration, the most common being to rock the incident light through carefully measured angles prior to strain, producing known changes in diffraction angle for constant groove spacing.

The apparatus required to make a diffraction grating determination of dynamic strain is shown in Fig.2.2 for a symmetrical free-flight impact experiment. Fig.2.3 shows the output of each of the photomultiplier tubes from the two first-order diffraction images from a 30,720

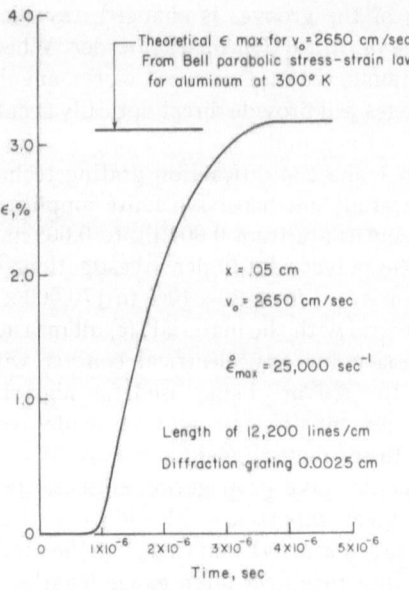

Fig. 2.1. An experimental diffraction grating strain vs time measurement at 0.05 cm from the impact face in a symmetrical free-flight impact experiment

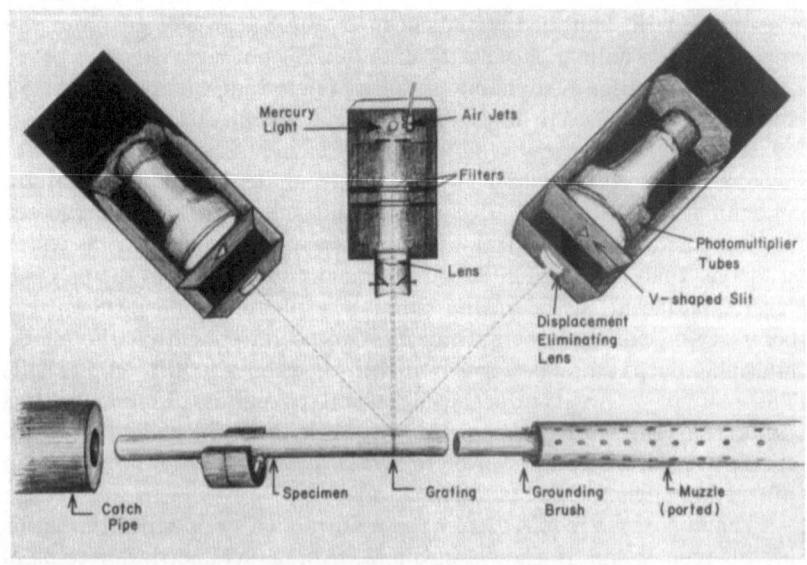

Fig. 2.2. The experimental apparatus for diffraction grating measurement of finite strain and surface angle in a free-flight symmetrical impact experiment

lines per inch, 0.005 in. long diffraction grating ruled at one-half diameter from the impact face.

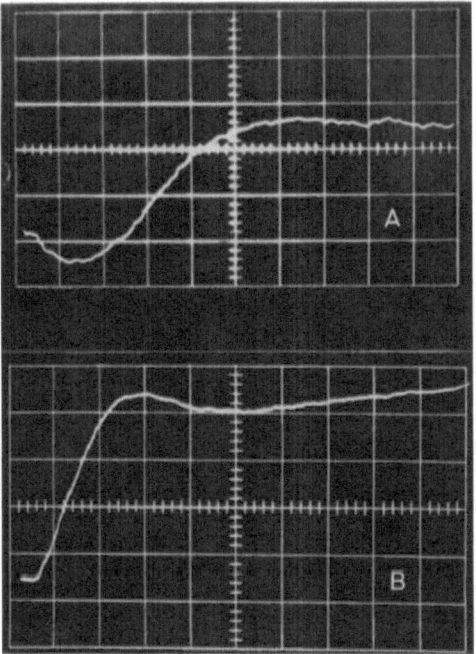

Fig. 2.3. The photomultiplier voltage outputs for a symmetrical free-flight impact experiment. The measurement was made at 1.26 cm from the impact face on a 25.4 cm long, 2.5 cm diameter aluminum specimen. Each small division is one microsecond

Each division in Fig. 2.3 is a microsecond. Through the calibration procedure these photomultiplier tube voltage outputs of Fig. 2.3 may be converted to the angles β_1 and β_2. For normally incident light the angle of diffraction θ_0 is given by Eq. (2.6),

$$\text{Sin } \theta_0 = \frac{n \lambda}{a} = n \lambda m_0 \tag{2.6}$$

where n is the diffraction order, λ the wave length of the monochromatic light employed, a the groove spacing and m_0 the initial number of lines per inch of the diffraction grating. During wave propagation the surface of the specimen does not necessarily remain perpendicular to the light source as the wave propagates through the point under observation. The angle of diffraction θ_n of Eq. (2.6) becomes Eq. (2.7), where α is the angle between the incident light and the normal to the surface.

$$\text{Sin } \theta_n - \text{Sin } \alpha = n \lambda m. \tag{2.7}$$

The number of lines per inch in Eq. (2.7) has been written as m rather than as m_0 to indicate that the diffraction grating spacing has been changed by the introduction of a deformation. The relation between m and m_0 is given by Eq. (2.8) for a compression strain ε.

$$m = \frac{m_0}{1-\varepsilon} \tag{2.8}$$

or

$$\varepsilon = 1 - \frac{m_0}{m}. \tag{2.9}$$

Writing θ_n in terms of the initial angle of diffraction and the two angular changes of the diffraction images β_1 and β_2 in Eq. (2.7) for each of two diffraction images, one may determine the instantaneous surface angle α of the wave front as Eq. (2.10) and the strain as Eq. (2.11).

$$\tan \alpha = \frac{\sin\left(\frac{\beta_1+\beta_2}{2}\right)\cos\left(\frac{\beta_1-\beta_2}{2}+\theta_0\right)}{1+\cos\left(\frac{\beta_1+\beta_2}{2}\right)\cos\left(\frac{\beta_1-\beta_2}{2}+\theta_0\right)} \tag{2.10}$$

$$\varepsilon = 1 - \frac{\sin\theta_0}{\sin\left(\frac{\beta_1-\beta_2}{2}+\theta_0\right)\cos\left(\frac{\beta_1+\beta_2}{2}-\alpha\right)}. \tag{2.11}$$

When the surface angle α is sufficiently small and strains are below 5% Eqs. (2.10) and (2.11) may be written as

$$\varepsilon = \frac{\beta_1-\beta_2}{2}\operatorname{Cot}\theta_0 \tag{2.12}$$

$$\alpha = \frac{\beta_1+\beta_2}{2}\frac{\operatorname{Cos}\theta_0}{1+\operatorname{Cos}\theta_0}. \tag{2.13}$$

The lathe ruling engine which is capable of producing 35,000 lines per inch rulings is shown in Fig. 2.4. A typical 0.005 in. long, 30,720 lines per inch diffraction grating produced on this cylindrical ruling engine is shown in Fig. 2.5. A first-order diffraction image from such a 0.005 in. long 30,720 lines per inch cylindrical diffraction grating for normally incident 5,461 λ monochromatic light is shown in Fig. 2.6. In Fig. 2.7 is shown a microphotograph of a cylindrical diffraction grating before and after a 3% compression strain. The strain-time and surface angle-time curves for the photomultiplier outputs shown in Fig. 2.3 are calculated as strain ε and surface angle α from Eqs. (2.12) and (2.13) in Fig. 2.8.

    In the symmetrical free-flight impact experiment shown diagrammatically in Fig. 2.9, a 0.005 in. long diffraction grating is ruled on the initially stationary specimen at prescribed distances from the impact face. The initially moving specimen is ejected from a gas gun whose muzzle is vented with ports as shown in Fig. 2.2 to produce a constant velocity impact during the dynamic deformation. The two specimens are optically aligned prior to impact. This alignment is checked at the instant of impact by a spark photograph of duration of less than one microsecond occurring at a time of within one-half microsecond of impact.

Fig. 2.4. The 35,000 lines per inch cylindrical ruling engine lathe

In addition to the V-shaped slits and corresponding photomultipliers of Fig. 2.9 for measuring diffraction angles, beam splitters, rectangular slits, and two additional photomultipliers are shown. These additional measurements are necessary to determine variations in the initial calibrations when large strains are measured (above $4^0/_0$ in Al polycrystals). For lower strains these additions are not required as the initial calibration is preserved. This change of initial calibration arises in light intensity changes produced by a roughening of the original polished surface adjacent to the grating. This phenomenon and the measurement

Fig. 2.5. A photomicrograph of a 0.005 in. long, 30,720 lines per inch diffraction grating. The spacing between threads is 0.0000325 in

of large strain, when it is present, have been described in considerable detail in a recent paper (BELL, 1967a). The air shock produced by the escaping gas just prior to impact has been found to provide a very accurate measure of dynamic specimen alignment (BELL, 1960b). Such a spark photograph is shown in Fig. 2.10 for a low velocity impact and in

Fig. 2.11 for a moderate impact velocity in which the striking specimen has an initial velocity of 4,060 cm/sec (in the symmetrical free-flight impact the impact velocity is one-half the striking specimen velocity v_h;

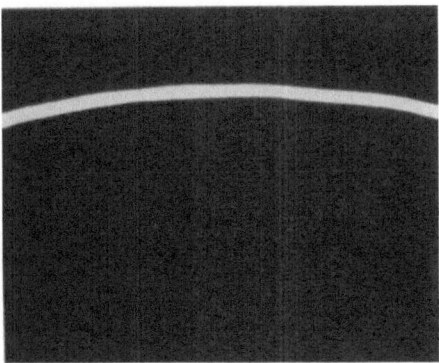

Fig. 2.6. A first-order diffraction image from a 0.005 in. long, 30,720 lines per inch cylindrical diffraction grating for normally incident 5,461 λ monochromatic light

Fig. 2.7. An 8,300 lines per inch diffraction grating before and after a 3% compression strain

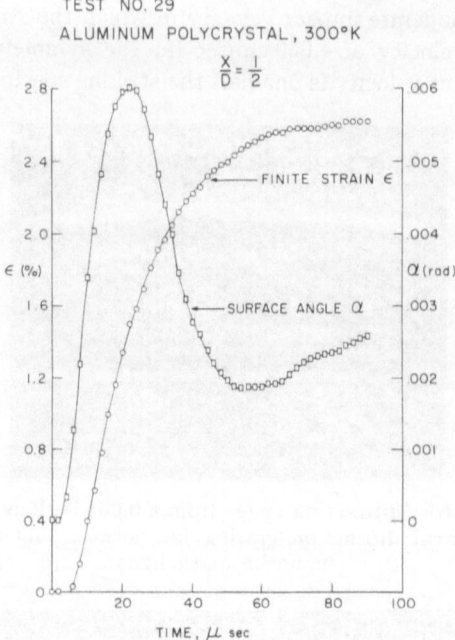

Fig. 2.8. Strain-time and surface angle-time data calculated from diffraction grating measurements

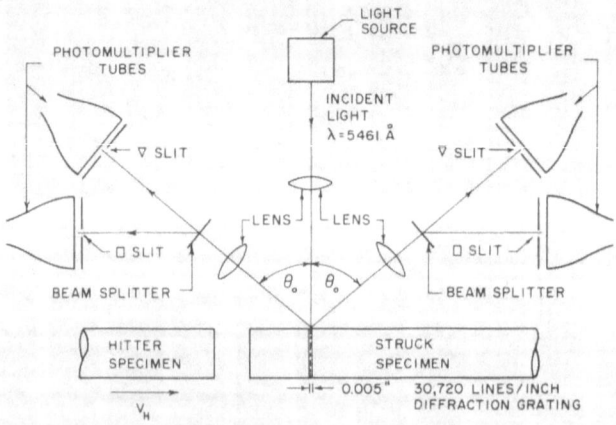

Fig. 2.9. A schematic diagram of diffraction grating apparatus with rectangular slits to monitor the calibration at large strains

i.e., $v_0 = 2{,}030$ cm/sec in Fig. 2.11). The air shock from deliberately misaligned specimens is shown in Fig. 2.12.

In Fig. 2.13 is shown the air shock for a relatively high striking specimen velocity of 15,250 cm/sec (BELL, 1967a). This spark photo-

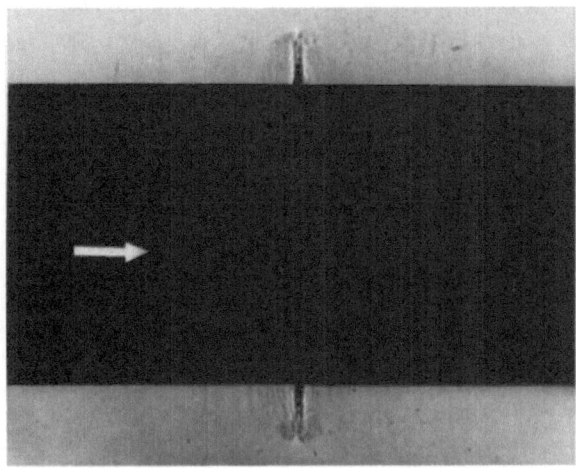

Fig. 2.10. A spark photograph of two 2.50 cm diameter specimens colliding axially at a low impact velocity

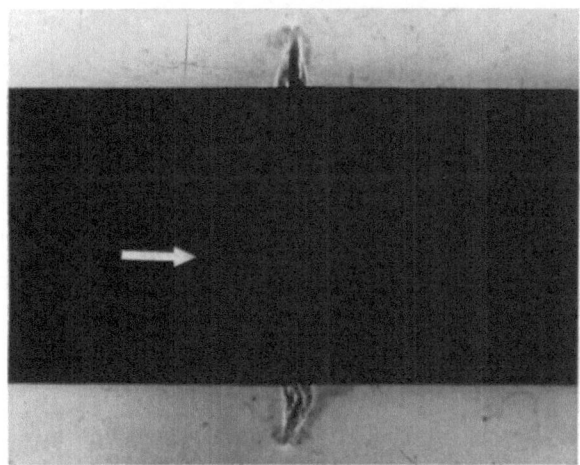

Fig. 2.11. A spark photograph of two 2.50 cm diameter specimens colliding axially at a moderate impact velocity

graph was made at a time of 9 microseconds after impact. The shock waves from the elastic wave front, which had progressed nearly 5 cm at the time of this photograph, may be seen as the triangular front in Fig. 2.13.

2*

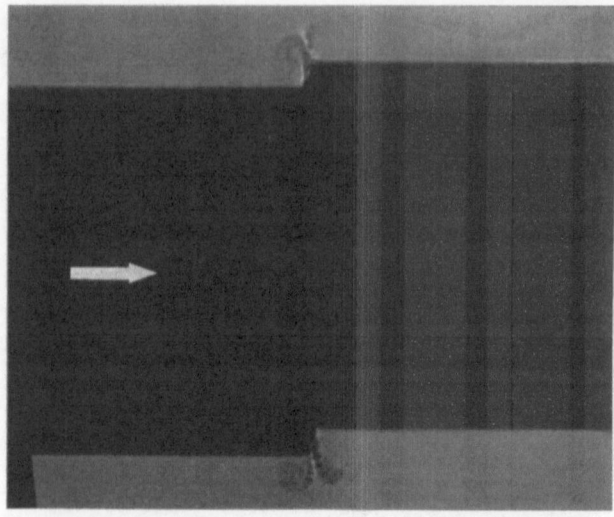

Fig. 2.12. A spark photograph of two deliberately misaligned 2.50 cm diameter specimens colliding axially at a moderate velocity

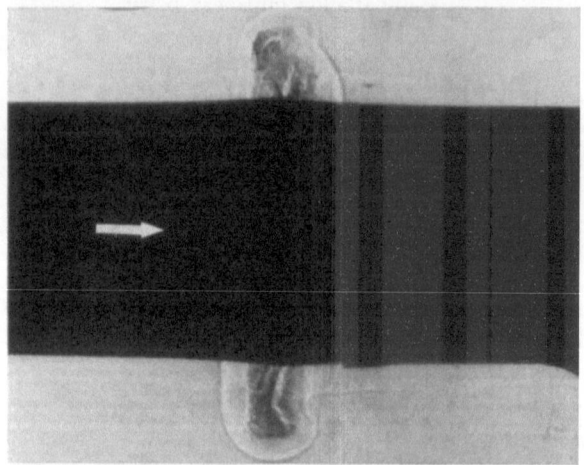

Fig. 2.13. A spark photograph of two 2.13 cm diameter specimens for a hitter velocity of 15,250 cm/sec, 9 microseconds after impact. Note the initial flaring of the specimen which has occurred before any reflection from the finite end of the specimen

The air shock of Fig. 2.13 is of sufficient magnitude to seriously interfere with the diffraction grating optics. All measurements at such high impact velocities are, therefore, conducted in vacuum chambers with optical measurements mad through quartz windows. Quartz ports are also employed in the side walls of furnaces for diffraction grating strain determinations at high ambient temperatures. Finite amplitude wave propagation has been studied in this manner up to within 20° of the melting point (BELL, 1962b, 1963a).

Earlier in the present chapter, it was stated that strain-time data in symmetrical free-flight impact experiments provided constant wave speeds for measurements at one or more diameters from the impact face. As illustrative of this experimental fact, the averaged data of 59 individual symmetrical free-flight experiments in long specimens of completely annealed fine grain commercial purity aluminum polycrystals are shown in Fig. 2.14.

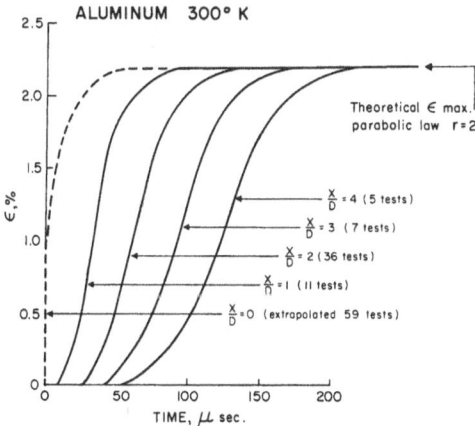

Fig. 2.14. Averaged strain-time data for 59 tests for an impact velocity $v_0 = 2,030$ cm/sec. The specimens are 25.4 cm long, with a diameter, $D = 2.50$ cm. X denotes the distance from the impact face

The distances from the impact face for these experiments were equal to spacings at 1, 2, 3, and 4 diameters. That wave speeds for each strain have different constant values is immediately apparent. The introduction of these constant wave speeds $C_p(\varepsilon)$ into the integral of Eq. (2.4) provides a relation between particle velocity $\dot{u}$ and finite strain ε. The impact velocity for these experiments of Fig. 2.14 was 2,030 cm/sec. The measured wave speeds from these data provide a maximum strain corresponding to this impact velocity of $v_0 = 2,030$ cm/sec of 2.19% which agrees with the total averaged maximum of $\varepsilon_{max} = 2.19\%$ shown in Fig. 2.14. (The actual experimental averages are $\varepsilon_{max} = 2.21\%$ for the 11 tests at $X/D = 1$, $\varepsilon_{max} = 2.19\%$ for the 36 tests at $X/D = 2$, $\varepsilon_{max} = 2.17\%$ for the 7 tests at $X/D = 3$, and $\varepsilon_{max} = 2.17\%$ for the 5 tests at $X/D = 4$.)

The experimental strain-time data of Fig. 2.14 are illustrative of diffraction grating measurements at one impact velocity. Similar data have been obtained for hitter velocities from less than 100 cm/sec to 15,250 cm/sec. For completely annealed aluminum at room temperature, this highest impact velocity corresponds to a finite amplitude wave front with a maximum initial strain of nearly 12%. In completely annealed polycrystalline aluminum, at each of the many impact velocities considered in this range, wave speeds are found not only to be constant at each strain amplitude, but the wave speeds for a given strain have the same numerical values independent of the strain rate or of the maximum finite strain amplitude of the wave front. At each impact velocity considered a large number of measurements had to be made so as to obtain averaged strain-time data at a sufficient number of positions along the specimen to determine both the constancy of wave speeds and their numerical value at each finite strain amplitude. From the analysis of over 400 individual symmetrical free-flight impact experiments in completely annealed polycrystalline aluminum at room temperature, it was discovered experimentally that the wave speeds $C_p(\varepsilon)$ were proportional to $\varepsilon^{-1/4}$. Upon substituting this empirical wave speed relation into the integral of Eq. (2.5) a parabolic governing stress-strain function, Eq. (2.14), was obtained.

$$\sigma = \beta \varepsilon^{1/2}. \tag{2.14}$$

For completely annealed aluminum at room temperature the parabola coefficient is $\beta = 5.6 \times 10^4$ psi or $\beta = 39.4$ kg/mm². (In view of the large number of experiments considered, to minimize error the data described in this monograph are given in the units of the experiment as performed or reported in the literature.) The original averages from which Eq. (2.14) was empirically established included approximately 150 measurements (BELL, 1960b). A later averaging of 450 experiments for a wider range of impact velocities and distances from the impact face provided the same parabola coefficient as the earlier determination (BELL, 1961b).

The substitution of the empirical wave speed $C_p(\varepsilon)$ into Eq. (2.4) provides the relation between particle velocity and strain of Eq. (2.15).

$$\dot{u}^2 = \frac{8}{9} \frac{\beta \varepsilon^{3/2}}{\varrho_0}. \tag{2.15}$$

If $\dot{u}$ is chosen as the impact velocity, v_0, the strain, ε, should have the maximum value $\varepsilon_{\max}$ and Eq. (2.15) becomes

$$v_0^2 = \frac{8}{9} \frac{\beta}{\varrho_0} \varepsilon_{\max}^{3/2}. \tag{2.16}$$

For a given impact velocity, the maximum measured finite strain beyond the first diameter has a constant value. It is thus a simple matter to intercompare the relation between particle velocity and a maximum strain from the same diffraction grating measurements from which the wave speeds $C_p(\varepsilon)$ were established. The average maximum strain throughout the entire range of impact velocities was found to provide a very close agreement with prediction from Eq. (2.16). Such a maximum strain plateau for the impact velocity of $v_0 = 2{,}030$ cm/sec was shown in Fig. 2.14. As may be seen in Fig. 2.14, the higher strains are delayed in traversing the first diameter so that only the lower finite strains provide wave speed data based upon zero impact time. The strain $\bar{\varepsilon}$ at which the initial delay begins has been found experimentally to be that of half the energy of deformation based upon the final deformation energy at $\varepsilon_{\max}$. Above this strain of the mean energy one may consider only traverse times between positions outside of the first diameter. In Figs. 2.15, 2.16, 2.17, and 2.18 are shown individual experiments in annealed aluminum polycrystals at the designated distance from the impact face and for the designated impact velocity. The relationship between wave speeds $C_p(\varepsilon)$ and ε is given by Eq. (2.17)

$$C_p{}^2(\varepsilon) = \frac{\beta}{2\varrho_0 \varepsilon^{1/2}} \qquad (2.17)$$

which, considering Eq. (2.14), may also be written as

$$C_p{}^2(\varepsilon) = \frac{\beta^2}{2\varrho_0 \sigma}. \qquad (2.18)$$

The solid lines in Figs. 2.15 through 2.18 are arrival times based upon zero time at the instant of collision [from Eq. (2.17)]. Also shown are the theoretical maximum values determined from Eq. (2.16) from the measured impact velocities v_0. In each symmetrical free-flight impact experiment, the velocity of the initially moving specimen just prior to impact is determined very accurately by means of an electronic chronograph. In a symmetrical free-flight collision of identical specimens, of course, the impact velocity v_0 is one-half the striking specimen velocity v_h.

To establish that the particle velocity-strain relation applies throughout the rising portion of the wave front, it is necessary to measure the longitudinal particle velocity-time histories under the same range of conditions as those under which finite strain was determined by means of diffraction gratings. An optical technique was developed to obtain displacement-time histories at any desired point along the specimen (BELL, 1961a, b). This experimental technique is shown diagrammatically in Fig. 2.19.

The specimen is covered with a non-reflecting black coating on one side of the position of interest, and a reflecting white coating on the opposite side. These coatings were specially prepared in the writer's

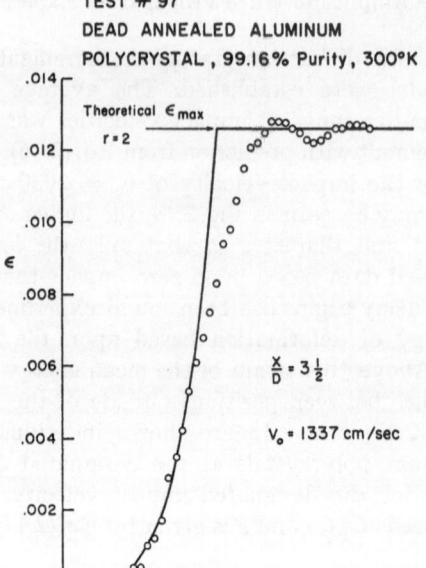

Fig. 2.15. A diffraction grating strain-time measurement (circles) compared with prediction (solid line) at $3\frac{1}{2}$ diameters from the impact face

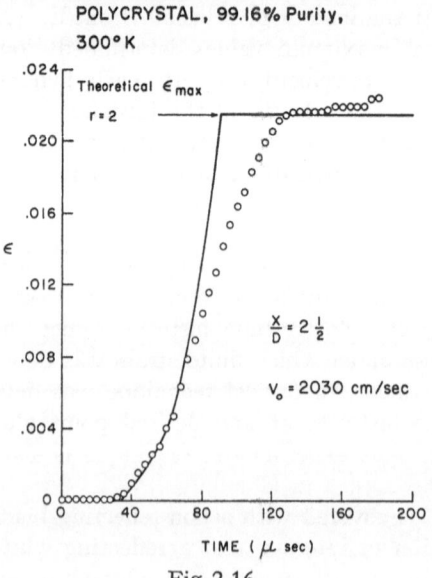

Fig. 2.16

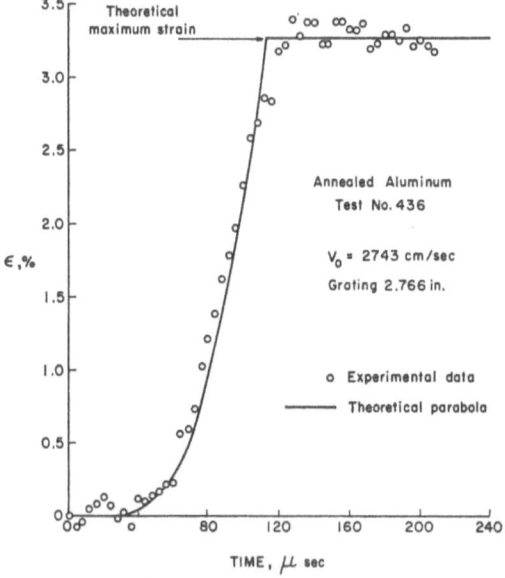

Fig. 2.17

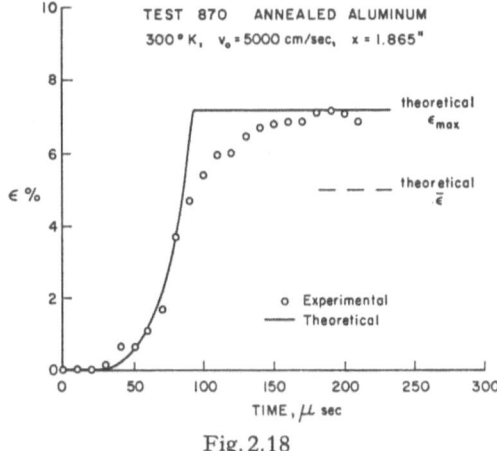

Fig. 2.18

laboratory to be able to withstand very large deformations without cracking. Calibrations were obtained in these experiments both by having specimens propagate through the uniform rectangular light field at precisely known impact velocities and by micrometer measurement of displacements as a function of light intensity. From the particle velocity relation of Eq. (2.15) the experimentally determined finite strain-time data could be used to provide a predicted displacement-time history at any point along the specimen for any given impact velocity. An example of four such measurements is shown in Fig. 2.20 for the same impact velocity; i.e., $v_0 = 2,030$ cm/sec, as the experimental data of Fig. 2.14.

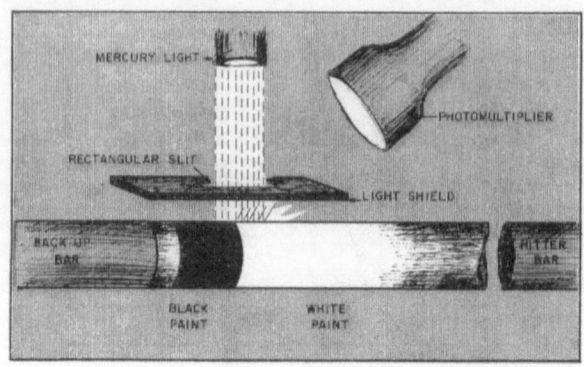

OPTICAL DISPLACEMENT EXPERIMENT

Fig. 2.19

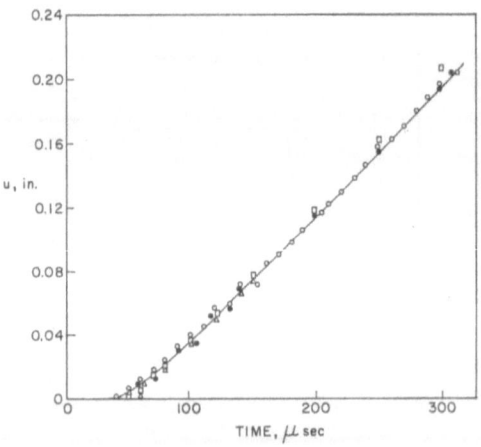

Fig. 2.20. Four displacement-time measurements for an impact velocity of $v_0 = 2,030$ cm/sec at 2 diameters from the impact face, compared with prediction from finite amplitude wave theory (solid line)

It is thus shown experimentally that the particle velocities $\dot{u}$ vs finite strain ε behavior of Eq. (2.15), required to establish the applicability of the finite amplitude wave theory Eq. (2.4), applies not only to the maximum values but also to all strain amplitudes during the rising portion of the non-linear wave front. This is of particular importance when one realizes that this wave front is distorted in traversing the three-dimensional first diameter.

Experimentally establishing that wave speeds are constant for given strain amplitudes, and that an invariable relation exists between particle velocity and strain, irrespective of the strain rate detail of the rising wave front, provides an experimental method for the systematic study of stress-strain functions in terms of Eq. (2.5) for all experimental situations in which it is demonstrated that these conditions hold.

To study finite amplitude wave propagation in symmetrical free-flight impact experiments at temperatures higher than room temperature, it is necessary to conduct the measurements inside furnaces (BELL, 1962b). Such furnaces are constructed with quartz windows through which the optical measurements of finite strain-time and particle velocity-time histories may be determined. In each instance, of course, one must establish that wave speeds, particle velocity, and strain are invariably related. In a series of experiments in completely annealed polycrystalline aluminum at numerous temperatures, from room temperature up to within 20° of the melting point $T_m = 932°$ K, the writer found that the finite amplitude wave theory was applicable in every instance (BELL, 1963a). The stress-strain function determined as a consequence from Eq. (2.5) was parabolic. Moreover, a comparison of parabola coefficients determined at various temperatures established a linear dependence with respect to the fractional melting point temperature T/T_m of Eq. (2.19) (BELL, 1963a).

$$\sigma = \beta(0)(1 - T/T_m)\varepsilon^{1/2} \qquad (2.19)$$

where $\beta(0)$ becomes the zero-point parabola coefficient which, for 99.16% purity fine grain dead annealed polycrystalline aluminum, has a numerical value of $\beta(0) = 58.0$ kg/mm². Experimental values of parabola coefficients β for dead annealed aluminum polycrystals as a function of T/T_m are shown in Fig. 2.21.

In 1962 BELL and WERNER described a series of diffraction grating finite amplitude wave propagation experiments in 99.9% purity fine-grained annealed copper polycrystals. This original series of experiments in copper was performed at two impact velocities. Results were obtained similar in every respect to those obtained in completely annealed polycrystalline aluminum. Wave speeds were constant for given strain amplitudes, and the theoretically predicted relation of Eq. (2.4) was found to be applicable. The substitution of the constant wave speeds $C_p(\varepsilon)$ into Eq. (2.5) also provided the parabolic stress-strain function of Eq. (2.14) with a room temperature parabola coefficient of 48.2 kg/mm². The melting point of copper is 1,358° K rather than the 932° K for aluminum. A substitution of this different fractional melting point temperature T/T_m into Eq. (2.19) provided a zero-point parabola cofficient of 61.7 kg/mm² which may be compared with the value of 58.0 kg/mm² for aluminum and with an identical value of 58.0 kg/mm² obtained

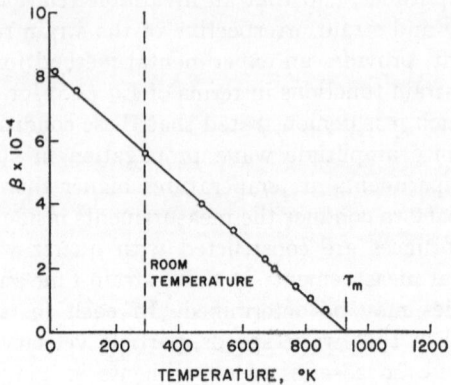

Fig. 2.21. Experimental parabola coefficients from 63 commercial purity polycrystalline aluminum free-flight impact experiments above room temperature

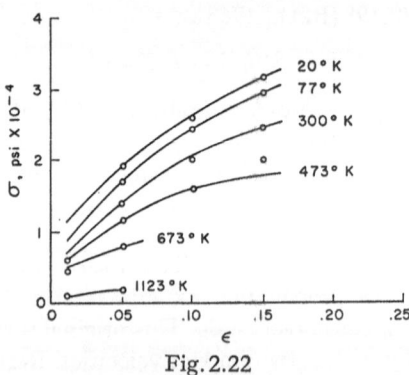

Fig. 2.22

from CARREKER's (1957) quasi-static polycrystalline data in annealed polycrystalline silver. These "detrued" silver data are compared in Fig. 2.22 at ambient temperature with prediction from aluminum for $\beta(0) = 58.0$ kg/mm² (BELL, 1963a).

From these aluminum, copper, and silver data it was first thought that $\beta(0)$ was a universal constant. Further evidence supporting this original conjecture was provided by a comparison of the quasi-static annealed aluminium polycrystalline data of CARREKER and HIBBARD (1957) and that of CARREKER (1957) in annealed silver at 20° K. These data coincided in the manner to be expected if $\beta(0)$ were a universal constant. This agreement is shown in Fig. 2.23 in which these data are compared with the parabola for $\beta(0) = 58.0$ kg/mm² (BELL, 1963a).

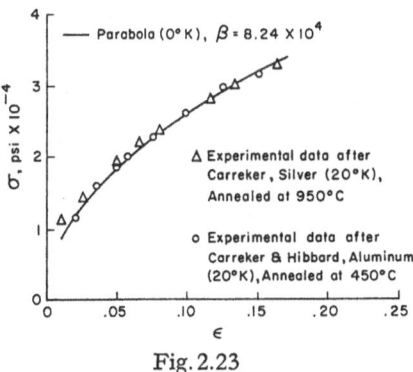

Fig. 2.23

Further support for this conjecture had been obtained earlier when these finite amplitude wave propagation polycrystalline stress-strain functions were compared with stage III[1] face-centered cubic copper single crystal data of numerous experimentalists and with the high purity aluminum single crystal data of POND and HARRISON (1958) at room temperature and of NOGGLE and KOEHLER (1957) at 4.2° K (BELL, 1961b, 1965a). These latter single crystal correlations are discussed in Chapter IV.

As the finite amplitude wave propagation of additional crystalline solids, such as zinc, magnesium, lead, nickel, α-brass, etc. were studied, it became apparent that although the zero-point parabola coefficient $\beta(0)$ was the same for some solids, this was not always the experimental situation. A comparison of several hundred stage III single crystal parabola coefficients described by the writer in 1964—1965, revealed that zero-point parabola coefficients were empirically interrelated in all instances in terms of an empirical multiplying factor $(3/2)^{3n/2}$ where $n = -1, 0, 1, 2, 3$ (BELL, 1965a). A major study was therefore undertaken to determine the origin of this discrete empirical factor.

From these single crystal studies, aluminum, silver, and gold were found to share, in general, common averaged parabola coefficients; nickel and iron shared common averaged parabola coefficients of a

---

[1] See Chapter IV for a definition of deformation stage for resolved stress-strain functions in single crystals.

much higher value. High purity copper had an intermediate averaged zero-point single crystal parabola coefficient value, while averaged values for lead lay far below. All of these zero-point parabola coefficients were expressible, one in terms of the other, by means of the empirical factor $(3/2)^{3n/2}$ (BELL, 1965a).

It was found, furthermore, that the same crystalline solids, under different conditions of purity and ambient temperature might have different values of n. A comparison of a wide variety of parameters such as elastic moduli, melting points, mass densities, lattice spacings, etc., among the elements for which parabola coefficients had been experimentally obtained revealed that the important parameter was the zero-point isotropic linear elastic shear modulus $\mu(0)$. The intercomparison of single crystal stage III zero-point parabola coefficients of high purity aluminum, copper, lead, and nickel at very low temperatures revealed that they indeed did fall precisely into the ratios of their respective zero-point isotropic shear moduli.

In terms of the ratios of the aggregate theory, a comparison between zero-point polycrystalline parabola coefficients obtained in finite amplitude wave studies, and stage III single crystal zero-point parabola coefficients, resulted in the discovery that all of these data for the many metals of several crystal types, including face-centered cubic, body-centered cubic, orthorhombic, and hexagonal were governed by a parabolic stress-strain function for which $\beta(0) = (2/3)^{r/2}\mu(0) B_0$. Substituting this value of $\beta(0)$ into Eq. (2.19) furnishes the polycrystalline stress-strain function:

$$\sigma = (2/3)^{r/2} \mu(0) B_0 (1 - T/T_m) \varepsilon^{1/2} \qquad (2.20)$$

where $\mu(0)$ is the zero-point isotropic linear elastic shear modulus; B_0 is a dimensionless universal constant having the value $B_0 = 0.0280$; and r is an integral index designating the large deformation mode ($r = 1, 2, 3, 4, ---$). In view of the introduction of experimental evidence (Chapter VI) establishing multiple-valued elastic moduli, the value of $\mu(0)$ in Eq. (2.20) is referred to as the stable modulus, experimentally associated with a near-zero stress state.

It will be the purpose of the next several chapters of this monograph to describe the nearly 1,000 quasi-static uniaxial stress experiments in 19 elements and 5 binary combinations which demonstrate the wide and remarkable generality of this stress-strain function discovered from the writer's dynamic wave propagation studies. This stress-strain function is known to be applicable over the entire temperature scale from 4° K to within 20° of the melting point in some crystalline solids. It is known experimentally to apply for strain rates from $\dot{\varepsilon} = 10^{-9} \sec^{-1}$ to $\dot{\varepsilon} = 10^4 \sec^{-1}$ for a ratio of 10^{13}. It is applicable to polycrystals and single crystals of a wide variety of purities and it is applicable to numerous binary combinations, including several metal alloys. All of the parameters of Eq. (2.20) are known except the integral index r. As will be shown in succeeding chapters, it is now known that the value of this discrete index depends upon primary variables such as temperature, loading history, etc. for a given crystalline solid.

Diffraction grating measurements of strain during finite amplitude wave propagation have been carried out in aluminium (BELL, 1956b, 1960b, 1961, 1962a, 1963a, 1967a), copper (BELL and WERNER, 1962), lead (SPERRAZZA, 1961b, 1962a, b), magnesium (CONN, 1959; BELL and CONN, 1968), zinc (BELL, 1968), and α-brass (HARTMAN, 1967) of different purities. Dynamic single crystal strain measurements have been made in high purity aluminum (GILLICH, 1964, 1967) and high purity copper (BELL and GILLICH, 1968). Wave speeds from strain-time measurements beyond the first diametral length of specimens also have been determined in nickel (FILBEY, 1963) and 24 S.T. aluminum alloy (CONN, 1965b) using an experimental technique which provides only a relative measurement of wave speeds. In each instance, when the observed wave speeds were introduced into the integral relationship of Eq. (2.5), a parabolic stress-strain function was obtained. The measured parabola coefficients thus obtained were expressible in terms of the generalized relation of Eq. (2.20).

The simplest manner of presenting examples of these experimental results is to compare measured strain-time data of the non-linear wave fronts, with prediction from the finite amplitude wave theory in terms of Eq. (2.21) and (2.22) for a parabolic stress-strain function, rather than to describe all of the wave propagation details in each instance. These equations are obtained from Eqs. (2.15) and (2.17) when the zero-point parabola coefficient of Eq. (2.20) is introduced:

$$\dot{u}^2 = \frac{8}{9} \times \frac{\left(\frac{2}{3}\right)^{r/2} \mu\,(0)\, B_0\,(1 - T/T_m)\, \varepsilon^{3/2}}{\varrho_0} \tag{2.21}$$

$$C_p{}^2(\varepsilon) = \frac{\left(\frac{2}{3}\right)^{r/2} \mu\,(0)\, B_0\,(1 - T/T_m)}{2\varrho_0\,\varepsilon^{1/2}}. \tag{2.22}$$

A similar introduction of the zero-point parabola coefficient of Eq. (2.20) into Eq. (2.18) furnishes:

$$C_p{}^2(\varepsilon) = \frac{\left(\frac{2}{3}\right)^{r} \mu^2\,(0)\, B_0{}^2\,(1 - T/T_m)^2}{2\varrho_0\,\sigma}. \tag{2.23}$$

It should be remembered that diffraction grating measurement of strain provides data in material coordinates and may best be correlated with finite amplitude wave theory in Lagrangian form. Therefore, the parabolic polycrystalline governing stress-strain function, Eq.(2.20), refers to the undeformed state of the material; i.e., it is a "nominal" or "engineering" stress-strain function. The deformation measure is, of course, arbitrary. Since the experimental data have shown the existence of parabolicity in nominal form, "true" stress-strain functions, referring to the deformed state of the material, obviously do not occur in such a parabolic form. It is an *experimental* fact, as is shown in later chapters, that the stress and strain ratios of the TAYLOR (1938) aggregate theory apply between *nominal* polycrystalline stress-strain functions and resolved stage III single crystal stress-strain functions.

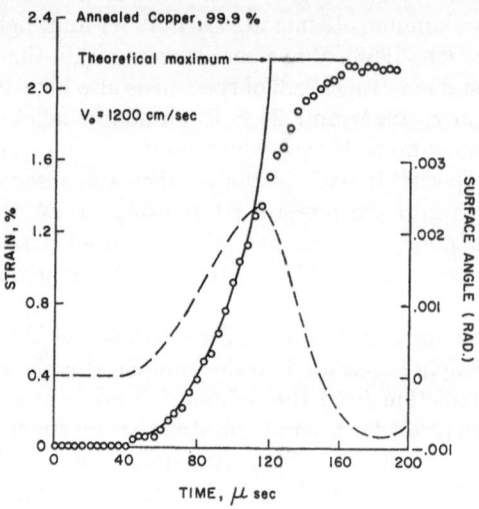

Fig. 2.24. A diffraction grating strain-time (circles) and surface angle-time (dashed line) measurement at 5.1 cm from the impact face, from the symmetrical free-flight impact of 25.4 cm long, 2.50 cm diameter cylinders, compared with prediction for $r = 4$. The impact velocity is 1,200 cm/sec

Diffraction grating strain-time data in annealed fine-grained 99.9% pure polycrystalline copper (circles) are compared in Figs. 2.24 and 2.25 with finite amplitude wave propagation theory prediction (solid lines) in terms of Eqs. (2.21) and (2.22) (BELL and WERNER, 1962).

Values of zero-point isotropic shear modulus $\mu(0)$ are tabulated below in Chapter V. The average experimental value for copper is 5,080 kg/mm² which is considerably higher than the 3,110 kg/mm² experimental value for aluminium. The mode index r for these copper data of Figs. 2.24 and 2.25 is $r = 4$ compared to the value of $r = 2$ for the 99.16% purity aluminum strain-time data of Fig. 2.14. As in aluminum, the strains below that of the mean deformation energy $\bar{\varepsilon}$ have arrival times corresponding to those for an infinite step at zero impact time with the upper portion of the strain-time curve distorted in traversing the first diameter where three dimensional effects predominate. In the BELL and WERNER (1962) experiments, the measured traverse times for strain-time data at the first diameter and beyond provided constant wave speeds at all strain levels. As may be seen in Figs. 2.24 and 2.25, the maximum strains, as well as initial wave speeds, agree with finite amplitude wave theory prediction from Eq. (2.22) for both impact velocities studied.

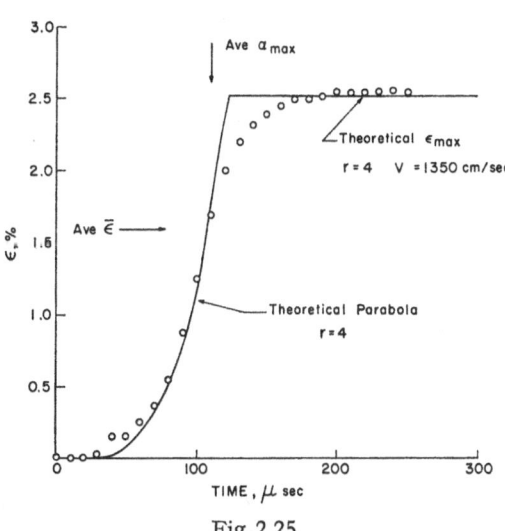

Fig. 2.25

Copper has a face-centered cubic crystal structure similar to that of aluminum. Polycrystalline zinc, on the other hand, has a hexagonal crystal structure. The purity of the fine-grained annealed zinc poly-crystals whose strain-time data is shown in Fig. 2.26 was chosen to be the same as aluminum; i.e., 99.2%.

Polycrystalline zinc in these experiments has the same large deformation mode index $r = 2$ as was found for the similar purity aluminum. The difference in parabola coefficient in Eq. (2.20) hence depends solely upon the ratio of the respective $\mu(0)$'s. The experimental $\mu(0)$ for zinc is 4,700 kg/mm²; for aluminum it is 3,110 kg/mm². In Eqs. (2.21) and (2.22) the difference in mass densities ϱ must also be included when aluminum and zinc wave speeds and strain-time data are compared. Such comparisons are shown in Fig. 2.26 for a series of zinc diffraction grating strain-time measurements at the designated impact velocity.

Assuming that first diameter development of aluminum and zinc are proportional to the ratio of their constant wave speeds beyond the first diameter, the experimental zinc strain-time data of Fig. 2.26 (solid lines) is compared with prediction from aluminum in terms of the differences of $\mu(0)$, T_m, and ϱ. Despite the fact that only four experiments in zinc are shown in Fig. 2.26 and that the comparison is between

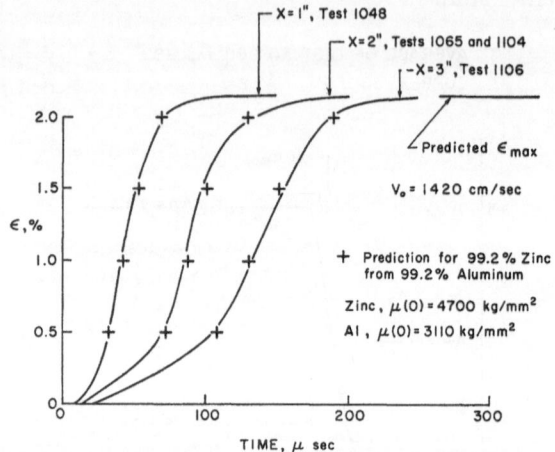

Fig. 2.26. Diffraction grating strain-time data for four experiments in annealed polycrystalline zinc (solid line) compared with prediction. Eq. (2.20) for $r = 2$ from aluminum (crosses)

a face-centered cubic metal and a hexagonal metal, the correlation in terms of the generalized parabolic stress-strain function of Eq. (2.20) is indeed remarkable.

The maximum strain for zinc in Fig. 2.26 is the same for each distance from the impact face x, shown. The impact velocity is one-half the measured striking specimen velocity in a symmetrical free-flight impact. Introducing the impact velocity $v_0 = 560$ in/sec for $\dot{u}$ in Eq. (2.21) provides the close agreement with experimental strain maxima shown in Fig. 2.26. Thus, one sees that polycrystalline zinc is very well described by the non-viscous finite amplitude wave theory.

Several years ago J. SPERRAZZA (1961, 1962, a, b) carried out in the writer's laboratory a series of finite amplitude wave propagation diffraction grating experiments in large grain multicrystals of high purity lead. Since the diffraction grating measurements were made on individual grains, it was necessary to average large amounts of experimental

strain-time data at different positions along the specimen to establish
that the finite amplitude wave theory did indeed apply to this crystalline
solid. In Fig. 2.27 diffraction grating strain-time data of 24 symmetrical
free-flight impact experiments at $X/D = 9/16, 3/2, 3$, and $9/2$ are scaled
at $X/D = 1$ for 99.99% pure lead mulitcrystals.

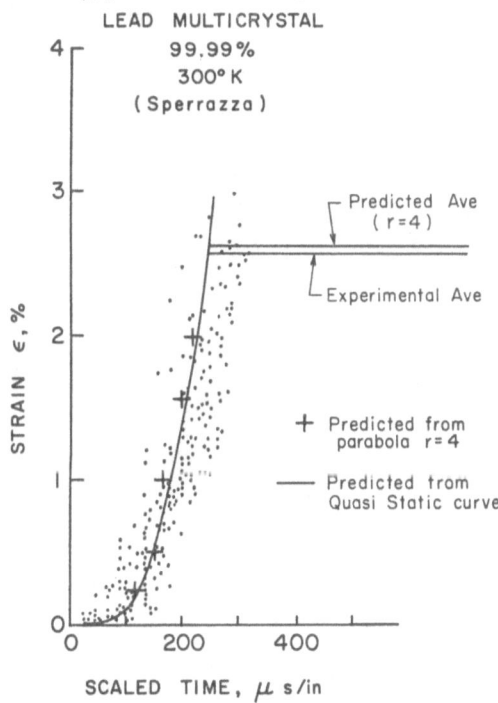

Fig. 2.27. 24 diffraction grating experiments in pure lead multicrystals at
X/D of 9/16, 3/2, 3, and 9/2, scaled to $X/D = 1$ by plotting as microseconds
per inch, compared with prediction (crosses) from the parabolic law, Eq. (2.20)
for $r = 4$, and with that from the quasi-static stress-strain function of
Fig. 2.28 (solid line)

These experimental data demonstrate that wave speeds and maximum strain in lead agree with prediction (crosses) from Eqs. (2.21) and
(2.22) for a mode index of $r = 4$. Here, $\mu(0)$ has the extremely low value
of $\mu(0) = 750$. The mass density ϱ is, of course, considerably higher
than that for aluminum, copper, or zinc. Because of the large grain size,
the most meaningful comparison for the lead polycrystalline data is
found in the averages.

Although at the time of SPERRAZZA's experiments (1961) the stress-
strain function of Eq. (2.20) had not yet been established, it may be
shown (BELL, 1963a) that if one introduces into Eq. (2.21) his averaged

maximum strain of 2.6% from experiments with an average impact velocity of 154 in./sec, one obtains a parabola coefficient $\beta(0) = 9.5 \text{ kg/mm}^2$. Since the melting point of lead is only $600° K$, the fractional melting point temperature T/T_m has the very large value of $T/T_m = 0.5$ for room temperature measurements at $300° K$. Nevertheless, this calculated polycrystalline parabola coefficient of 9.5 is in very close agreement with the value from $\beta(0) = (2/3)^{r/2} \mu(0) B_0$ of 9.36 kg/mm^2 predicted for a value of the mode index $r = 4$. As will be shown in Chapter IV, room temperature single crystal stage III parabola coefficients of 99.99% purity lead also have this same deformation mode index of $r = 4$ in the data of BOLLING, HAYS and WIEDERSICH (1962).

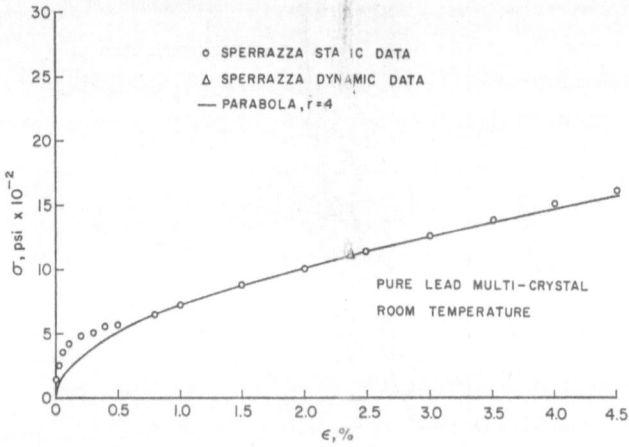

Fig. 2.28. Quasi-static uniaxial compression lead experiment (circles) compared with Eq. (2.20) for $r = 4$, (solid line). The triangle is determined from the dynamic data by Eq. (2.21)

SPERRAZZA (1961, 1962, a, b) provided a quasi-static polycrystalline uniaxial compression measurement in a lead multicrystal specimen of the same purity and grain size as the dynamic data. This writer (BELL, 1963a) has shown that the data from SPERRAZZA's lead quasi-static experiment was parabolic and, as is shown later in Chapter III, the initial deformation mode of this parabola is the same as that in the dynamic data; i.e., $r = 4$. A σ vs ε plot of this quasi-static SPERRAZZA experiment is shown in Fig. 2.28 where it is compared with the maximum stress and maximum strain obtained by SPERRAZZA, from averaging his finite amplitude wave propagation data in terms of Eqs. (2.4) and (2.5). Arrival times determined from the slopes of this quasi-static curve are shown in Fig. 2.28.

One of the most interesting of the polycrystalline finite amplitude wave propagation studies is a series of diffraction grating measurements

in fine-grained low purity polycrystalline magnesium performed by
CONN (1959) and by BELL and CONN (1968). What is unique about these
measurements is that the quasi-static uniaxial compression stress-
strain function is upward turning. If such a stress-strain function
governs finite amplitude wave propagation, then as WHITE and GRIFFIS
(1942) showed 25 years ago, one would expect the creation of shock
wave fronts similar to the situation for non-linear progressive waves in
a compressible gas. Such experimental quasi-static stress-strain curves
for annealed magnesium are shown in Fig. 2.29.

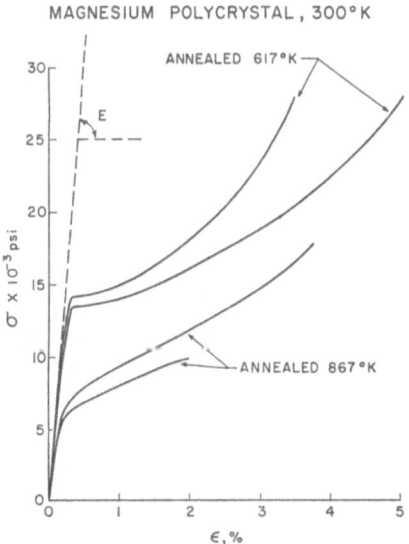

Fig. 2.29. Nominal compression stress-strain data in polycrystalline
magnesium

The elastic limit for the experiment in which the specimen was
annealed at 617° K is approximately 14,000 psi. Raising the annealing
temperature to 867° K resulted in the lowering of this elastic limit to
the minimum value obtainable, 6,000 psi. The presence of such relatively
high elastic limits creates dificulties in finite amplitude wave propagation
studies not only in the appearance of an instability in the separation
of the initial high speed elastic and slow moving plastic wave fronts,
but also in a decrease in the total time available for finite amplitude
wave measurement. Despite the upward turning quasi-static stress-
strain curve, shock wave fronts predicted by WHITE and GRIFFIS (1942)
were *not* observed in the diffraction grating measurements in symmetrical
free-flight impact experiments in polycrystalline magnesium. Instead,
as is shown in Fig. 2.30, where the average strain-time data of seven

experiments at one diameter are shown, the wave speeds and maximum strain are those predicted by the finite amplitude wave theory from Eqs. (2.16) and (2.17) for a deformation mode index, $r = 2$.

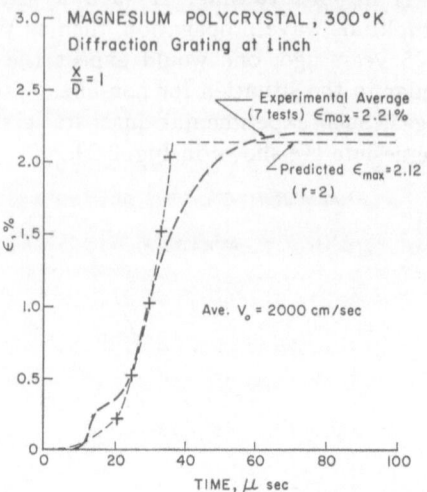

Fig. 2.30. The experimental average of 7 diffraction grating experiments in polycrystalline magnesium at $X/D = 1$, compared with prediction from Eq. (2.20)

For magnesium, $\mu(0) = 2{,}050 \text{ kg/mm}^2$ and $T_m = 924°$ K. The wave speed and maximum strain data in these magnesium experiments are described in detail by BELL and CONN (1968). The value of the maximum strain is independent of the initial elastic limit or the magnitude of the initial linear elastic wave front. This independence of maximum finite strain and the elastic limit, coupled with the fact that the upward turning quasi-static stress-strain curve obviously does not govern the finite amplitude wave propagation, demonstrates the generality of the parabolic stress-strain function of Eq. (2.20).

HARTMAN (1967) has performed in the writer's laboratory an extensive series of both finite amplitude wave propagation experiments and uniaxial quasi-static compression and tension experiments to determine whether or not the present writer's linearly temperature dependent parabolic generalized stress-strain function of Eq. (2.20) is applicable to a binary alloy. HARTMAN was particularly interested in choosing for his study a binary alloy for which the reported stacking fault energy was widely different from aluminum. He found that 70—30 α-brass fit these requirements. It has a $\mu(0) = 4{,}660 \text{ kg/mm}^2$ and a solidus melting point $T_m = 1{,}188°$ K, both of which differ from the value of aluminum. The mass density of the 70—30 α-brass also is very much higher than that of aluminum. Two of HARTMAN's (1967) nearly 40 finite amplitude wave propagation experiments using the same diffraction grating apparatus as that employed in aluminum, copper, lead, zinc, and magnesium studies described above are shown in Fig. 2.31.

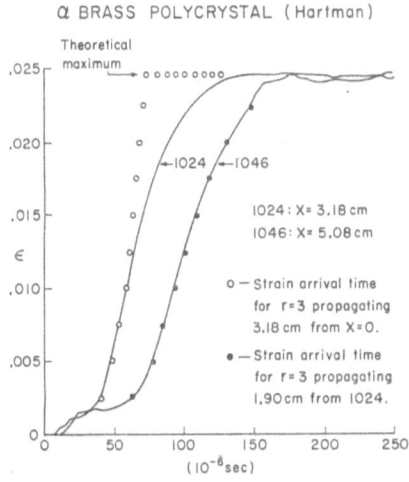

Fig. 2.31. Two diffraction grating experiments in 70—30 α-brass, compared with prediction, Eq. (2.20), assuming an infinite step at the impact face (open circles) and with predicted traverse time from Eq. (2.22) from one position to the other

These strain-time curves (solid lines) are compared with arrival time and maximum strain prediction from Eqs. (2.21) and (2.22) for a large deformation mode index, $r = 3$. As will be described in later chapters, HARTMAN (1967) found that α-brass is in close agreement in every respect with the linear temperature dependent parabolic generalization of Eq. (2.20). HARTMAN has found that this face-centered cubic crystalline solid has a deformation mode index r less predictable than the index for aluminum, in that, the value obtained in finite amplitude wave studies may vary from specimen to specimen depending upon small variations in the prior metallurgical treatment. Nevertheless, all of HARTMAN's data falls into one or another of the discrete groups of zero-point parabola coefficients.

FILBEY (1963) carried out experiments in low purity polycrystalline nickel which included both wave propagation studies in symmetrical free-flight impact and quasi-static compression experiments. These latter data are described in the next chapter. FILBEY's (1963) finite amplitude wave propagation experiments in nickel were not performed

with diffraction gratings so that it was not possible to determine whether
Eq. (2.16) was applicable. The strain-time measurements were made by
means of wire resistance strain gauges which, as GILLICH (1960) and
BELL (1960d) have shown, introduce errors both in arrival time and
maximum strain amplitude at a given X/D position. However, such
measurements do provide fairly accurate wave speed data for traverse
times between two positions in a symmetrical free-flight impact experiment. These wave speed measurements provided experimental agreement with one of the two conditions of the finite amplitude wave theory.
Since both of the conditions could not be checked experimentally, the
question of the applicability of the finite amplitude wave theory was
not completely answered with respect to FILBEY's (1963) nickel data.

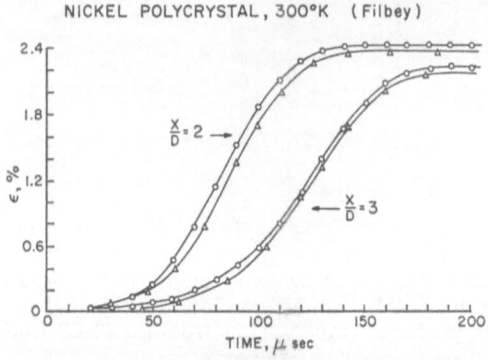

Fig. 2.32. Two experiments in polycrystalline nickel obtained with wire
resistance strain gauges for simultaneous measurements at $X/D = 2$ and
$X/D = 3$

In Fig. 2.32 are shown FILBEY's (1963) strain-time data which, with
the limitations indicated above, provide a zero-point parabola coefficient
$\beta(0)$ for nickel of 89 kg/mm² which may be compared with the value
determined from introducing $\mu(0) = 8,600$ kg/mm², $T_m = 1,725°$ K,
into Eq. (2.22) where the mode index $r = 5$. This calculated value of
$\beta(0)$ for $r = 5$ is 88 kg/mm². As will be shown in Chapter III, FILBEY's
quasi-static nickel test also provides the same parabola coefficient of
$r = 5$.

In all of the crystalline solids considered in this chapter, both conditions of the finite amplitude wave theory with respect to wave speeds,
particle velocity, and finite strain, were established experimentally
before a governing stress-strain function was determined. In FILBEY's
(1965) nickel data, only one of these conditions was established. Nevertheless, the observed wave speeds are found to be in agreement with
prediction from Eq. (2.22) for this crystalline solid.

Transition Velocities

Of the seven crystalline elements whose finite amplitude wave propagation behavior has been considered in this chapter, the solid most thoroughly studied by the writer is aluminum. Several hundred dynamic plasticity experiments have been performed in this element for the purpose of considering in great detail not only the applicability of the TAYLOR (1942), VON KARMAN (1942), RAKHMATULIN (1945) finite amplitude wave theory and the governing stress-strain function, but also the dependence of this behavior upon purity, grain size, specimen diameter, ambient temperature, impact velocity, and prior metallurgical treatment of the specimen. A very large number of experiments also have been performed by the writer to study the problems of non-linear wave initiation and growth in the immediate vicinity of the impact face (BELL, 1960c, 1961b, 1963b, 1965b), reflections from free and elastic boundaries (BELL, 1961a, b, c; 1963b, 1965b, 1966b) incremental waves in prestressed fields (BELL, 1951; BELL and STEIN, 1962), and the general problem of the interaction of loading wave fronts with both loading and unloading wave fronts propagating in opposite directions (BELL, 1961a, c).

Most of these studies are not of direct interest in the present monograph which deals primarily with the matter of constitutive relations. One aspect of these aluminum wave propagation studies, however, is of great significance in the present context. As the impact velocity was increased from very low to higher values, a series of critical velocities were observed and reported upon (BELL, 1960c, 1961b, 1962a, 1963b, 1965b, 1967a; BELL and SUCKLING, 1962). The governing parabolic stress-strain function, Eq. (2.20), with a single mode index r is operable throughout the entire rising portion of the strain-time fronts to the highest impact velocity that has been studied. For commercially pure dead-annealed aluminum in which the initial velocity transition behavior has been most thoroughly studied, the value of the mode index is $r = 2$. The observed transition phenomena are seen either as a major change in first diameter wave initiation or as a change in the specimen optical reflectivity as the particle velocity exceeds a specified value.

The first critical velocity of $v_0 = 1,478$ cm/sec (582 in./sec) was referred to by the writer in 1961 as the VON KARMAN critical velocity because the same value of velocity had been obtained by R. B. POND (private communication) from an integration of the slopes of quasi-static tension stress-strain function to the strain at the horizontal tangent in the manner earlier suggested by VON KARMAN (1942). In the writer's compression experiments the behavior below and above this first critical velocity may be seen from the comparison of the two diffraction grating experiments of Figs. 2.33 and 2.34, where the 30,720 lines/in. diffraction gratings were located at 1/8″ from the impact face. Below this impact velocity there is a delay of nearly 20 μsec in the initiation of the plastic deformation at the impact face. The maximum strain obtained is almost exactly the theoretical value predicted from the finite amplitude wave theory. This behavior is shown in Fig. 2.33 at an impact velocity of 1,360 cm/sec just below the critical value of 1,478 cm/sec.

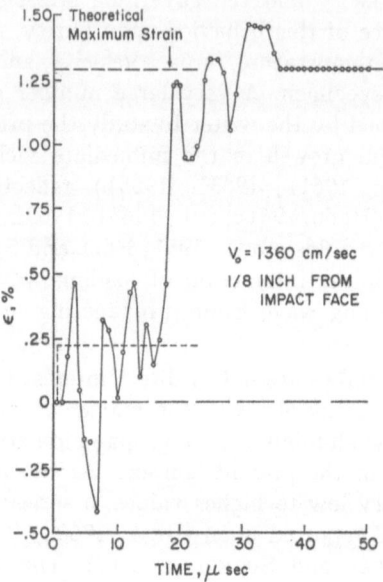

Fig. 2.33. A diffraction grating measurement of strain in polycrystalline aluminum just below the first transition velocity, showing the delay in the development of finite deformation. The lower dashed line corresponds to linear elastic behavior. Note the experimental strain rate of $\dot{\varepsilon} = 12{,}000$ sec^{-1} for this low impact velocity experiment

    The strain-time data of Fig. 2.34, on the other hand, at an impact velocity of $v_0 = 1{,}700$ cm/sec just above 1,478 cm/sec rises to the predicted maximum strain without the initial delay in the formation of the plastic deformation. This experiment, No. 265, of Fig. 2.34 also exhibits the initial two-wave structure at the impact face which has been the subject of extensive earlier study by the writer (BELL, 1961 a, b; 1962 a, 1963 a; BELL and WERNER, 1962). It should be noted particularly that below the critical velocity of 1,478 cm/sec there is no initial flaring or mushrooming of the first diameter of the specimen which is characteristic of behavior above this critical velocity. Introducing this critical velocity

$v_t = 1{,}478 \text{ cm/sec}$ into Eq. (2.21) for commercially pure aluminum, with a mode index $r = 2$, one obtains a transition strain $\varepsilon_t = 1.45\%$.

As the impact velocity for finite amplitude wave propagation in completely annealed aluminum polycrystals is increased above the first transition velocity of 1,478 cm/sec, no abnormalites are observed until a second critical value of $v_0 = 3{,}350 \text{ cm/sec}$ is reached. At this impact velocity, which in terms of the $r = 2$ mode index parabolic stress-strain function for this solid corresponds to a maximum strain from Eq. (2.21)

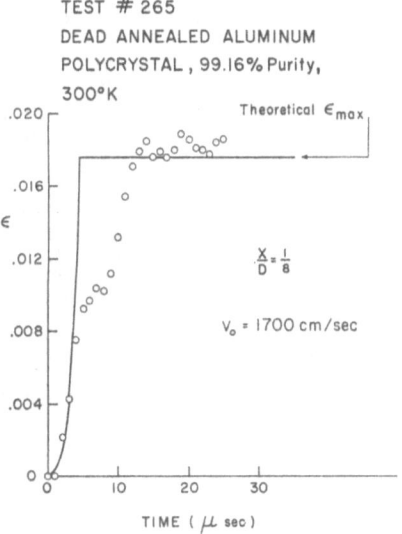

Fig. 2.34. A strain-time experiment at the same position near the impact face as in Fig. 2.33; just above the first transition velocity

of 4.2%, a new phenomenon inevitably is observed (BELL, 1967a). The surface of the aluminum specimen when strained dynamically above 4.2% suddenly becomes exceedingly mottled so that its optical reflectivity for diffraction grating measurement is radically changed. A few years ago when this sudden change in reflectivity was observed to nullify the original calibration of the diffraction grating, a modification of this optical method of measuring larger strain had to be introduced before larger amplitude strain fronts could be studied (BELL, 1966a, 1967a). The modification consisted of using beam splitters to project a portion of the light from each diffraction image to a second set of photomultiplier tubes with rectangular slits (see Fig. 2.9). The details of this modification of the original experiment have been described in a recent paper (BELL, 1967a) as has been the fact that this modification has allowed the accurate measurement of finite strain to as high as 17%.

By studying experimental strain-time data above and below the transition strain of 4.2%; i.e., impact velocity $v_0 = 3,350$ cm/sec, this value of the transition strain was very accurately determined.

A third transition velocity was first reported by the writer (BELL and SUCKLING, 1962; BELL, 1963b) to occur at an impact velocity $v_0 = 5,080$ cm/sec (2,000 in/sec) corresponding to a maximum strain for a mode index for aluminum of $r = 2$ of $\varepsilon_{max} = 7.5\%$. BELL and SUCKLING (1962) had designated this impact velocity as the hydrodynamic transition velocity. This was perhaps an unsatisfactory choice of terminology since subsequent research has revealed the existence of even higher critical velocities.

Above the impact velocity of 5,080 cm/sec finite amplitude wave initiation in the first one-half diameter of the specimen is radically altered as FILBEY (1961a, b) has shown from direct diffraction grating studies, although the finite amplitude wave propagation from one diameter onward is still governed by the parabolic stress-strain function of Eq. (2.20) as in all other velocity ranges. FILBEY (1961a, b; 1963) showed that above the third critical velocity, wave speeds depend only on strain in the first one-half diametral length of specimen. The governing stress-strain function close to the impact face calculated from these wave speeds was cubic rather than parabolic.

Piezo crystal measurements of impact face stress-time histories during symmetrical free-flight impact have shown (FILBEY, 1961a, b; BELL and SUCKLING, 1962; BELL, 1961b, 1962a) that during the first microsecond a very large initial stress occurs which immediately falls through an intermediate stress plateau, referred to as the dynamic overstress, to the maximum stress of the parabolic law. The maximum value of this initial stress, called by the writer the "initial peak stress" increases in a calculable manner (BELL, 1962a) until the critical velocity of 5,080 cm/sec is reached after which no further increase is observed as the impact velocity increases. This peak stress is empirically given by Eq. (2.24).

$$\sigma_p = \varrho_0 c_0 v_n \qquad (2.24)$$

where $c_0 = \sqrt{E/\varrho_0}$ and v_n is the critical velocity.

In an earlier paper the author (BELL, 1963b) showed that the experimental transition velocity of 5,080 cm/sec in completely annealed fine-grained commercial purity aluminum corresponded, in terms of the writer's generalized parabolic law, to a value of stress of 15,400 psi, which is identical with the tension dynamic ultimate strength published by CLARK (1953), and very close to the room temperature value of dynamic ultimate strength given by MANJOINE and NADAI (1940), CLARK and WOOD (1950), and other experimentalists. Above this value of transition velocity of 5,080 cm/sec in compression tests, corresponding to the dynamic ultimate strength of 15,400 psi in tension tests, the first diameter dynamic overstress of JOHNSON, WOOD, and CLARK (1953) disappears, as has been shown experimentally by FILBEY (1961a, b) and by the present author (BELL and SUCKLING, 1962; BELL, 1963b). The writer has shown (BELL, 1967a) for an impact velocity of 7,600 cm/sec (far above the third transition velocity of 5,080 cm/sec) these wave speeds of FILBEY's cubic law still govern the wave propagation more than four

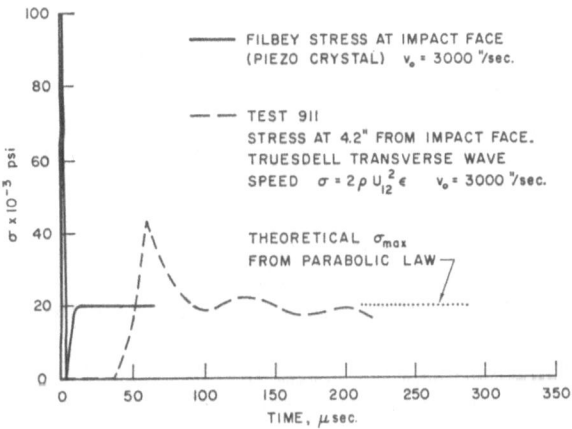

Fig. 2.35. A comparison of stress-time behavior at the impact face and at 4.2 in. from the impact face. Both sets of experimental data have a final stress agreeing with the parabolic stress-strain function of Eq. (2.20) for
$$r = 2$$

diameters from impact for the initial shock front, which precedes the conventional finite amplitude wave front. In earlier papers the writer has shown that the initial peak stress determined from piezo crystal measurements at the impact face has a maximum which is determinable empirically from Eq. (2.24) (BELL, 1961b, 1962a).

It was furthermore shown (BELL, 1962a) that all of the detail of the rising and falling portion of the peak stress, as well as the subsequent detail of the stress-time front, is *empirically* given by the wave speed relations of TRUESDELL (1961) in his theory of wave propagation in finite elasticity. These wave speed relations which interrelate stress, strain, and wave speed also have been shown (BELL, 1962a) to be capable of providing stress-time histories from measured strain-time and wave speed data for the FILBEY (1961a, b; 1963) weak shock fronts at distances exceeding 4 in. from the impact face, as may be seen in Fig. 2.35.

The third transition velocity of 5,080 cm/sec furnishes a maximum value of the initial peak stress, Eq. (2.24), of 100,000 psi, which is in close agreement with experiments at that velocity. A piezo crystal measurement of the peak stress at impact face is shown in Fig. 2.35. When the impact velocity is increased above 5,080 cm/sec (7,800 cm/sec was the maximum velocity studied in this series) one still obtains the same limiting value of 100,000 psi. This was the original reason for designating 5,080 cm/sec as the hydrodynamic transition velocity (BELL and SUCKLING, 1962). It also is important that this impact velocity of 5,080 cm/sec in completely annealed aluminum may be calculated from a knowledge of the elastic limit Y of the solid in terms of Eq. (2.25).

$$\varrho_0 v_t^2 = Y. \tag{2.25}$$

An additional property of dynamic plastic deformation above this transition velocity is the appearance of radical flaring or mushrooming of the first diameter of the specimen. An inspection of the spark photograph of Fig. 2.13 which was for an impact velocity of 7,600 cm/sec reveals that at 9 μsec, long before any reflected waves have returned from the finite ends of the specimens, this initial flaring already has occurred.

Transition phenomena have been introduced at this point in this monograph, which is primarily concerned with the generalized stress-strain function for the finite distortion of crystalline solids, in order to show that the three strains, $\varepsilon_t = 1.5^0/_0$, $\varepsilon_t = 4.2^0/_0$, and $\varepsilon_t = 7.5^0/_0$, corresponding to respective impact velocities of 1,478 cm/sec, 3,350 cm/sec, and 5,080 cm/sec, first observed by the writer several years ago in dynamic deformation, are the same strains at which quasi-static transitions from one deformation mode to another occur in the *quasi-static* uniaxial stress finite deformation of both single crystals and polycrystals.

Determination of Maximum Stress

Earlier in the present chapter it was shown that a comparison of the finite amplitude theory of non-linear wave propagation with experiment does not require the measurement of dynamic uniaxial stress. Nevertheless, an extensive series of experiments was performed, because of the fact that strain and particle velocity measurements are made upon the surface of the specimen, whereas end-face stress determinations represent an integration over the entire cylindrical specimen cross-section.

Of the many types of experiments by means of which the present writer has studied this phenomenon, three are relevant in the present context. These three types of experiments are shown in Fig. 2.36.

Of these three experiments, only the piezo crystal measurement is a symmetrical free-flight impact of identical specimens under conditions nearly equivalent to those in which the writer's finite amplitude wave propagation studies were made. The load bar experiment and the elastic-plastic boundary experiment introduce radial friction at the boundary since the hard bar remains in the infinitesimal elastic strain condition.

In Fig. 2.37 is shown an example of a piezo crystal experiment for an annealed aluminum polycrystal. One sees peak stress behavior during the first few μsec, and the subsequent stress-time behavior which provides the close agreement with prediction from the parabolic stress-strain function, Eq. (2.20), for $r = 2$.

PIEZO CRYSTAL EXPERIMENT

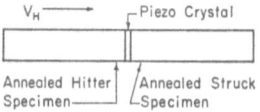

LOAD BAR EXPERIMENT

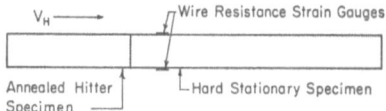

ELASTIC – PLASTIC BOUNDARY EXPERIMENT

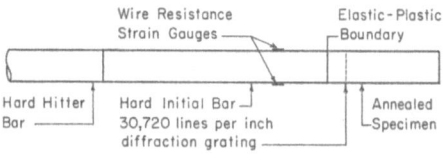

Fig. 2.36

Shown in Fig. 2.38 are two load bar experiments in annealed copper. In each instance there is evident a very close agreement between the observed maximum stress and the predicted stress from the finite amplitude wave theory, Eq. (2.20).

A similar measurement in aluminum is shown in Fig. 2.39. Here it may be seen that following an initial first diameter calculable dynamic overstress, the observed maximum stress in this very long annealed specimen at this relatively low impact velocity is in excellent agreement with prediction from the parabolic stress-strain function.

And, finally, in Fig. 2.40, an example of one of the many elastic-plastic boundary experiments of the present writer is shown for annealed polycrystalline aluminum at the same impact velocity as the load bar test of Fig. 2.39. Through a consideration of the common stress and particle velocity at the elastic-plastic interface, the maximum stress for the parabolic stress-strain function may be determined. An examination of Fig. 2.40 reveals that this is in fact the maximum stress observed after the initial calculable first diameter dynamic overstress.

These experiments of Figs. 2.37 through 2.40 reveal that the experimental maximum stress-impact velocity behavior is entirely consistent

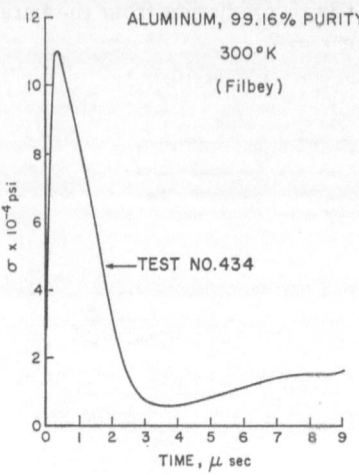

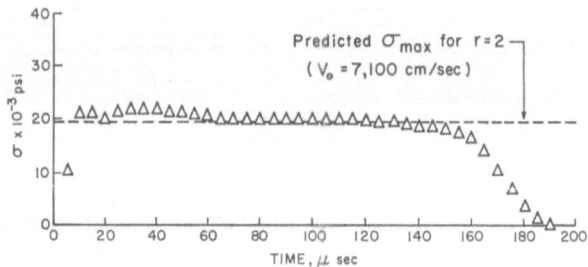

Fig. 2.37. Piezo crystal determination of stress in high velocity impact, showing the initial peak stress from the first microsecond and the subsequent maximum stress of the parabolic stress-strain function, Eq. (2.20)

with expectation in terms of the finite amplitude wave theory governed by the writer's generalized linearly temperature dependent parabolic stress-strain function. One is also assured by these measurements that the dynamic behavior extends over the entire cross-sectional area of the cylindrical specimens.

A number of additional experimental studies have been made to be certain that the finite amplitude deformation wave is diametrally uniform across the cylinder. These studies include the interaction of loading

waves with unloading waves from the free end of finite specimens
(BELL, 1961a, c); studies of plastic-elastic boundary loading wave
interaction, as well as the elastic-plastic boundary behavior described
above; and the study of symmetrical free-flight impact in hollow tubes
at 2.50 cm diameter with a 0.315 mm wall thickness (BELL, 1960b),

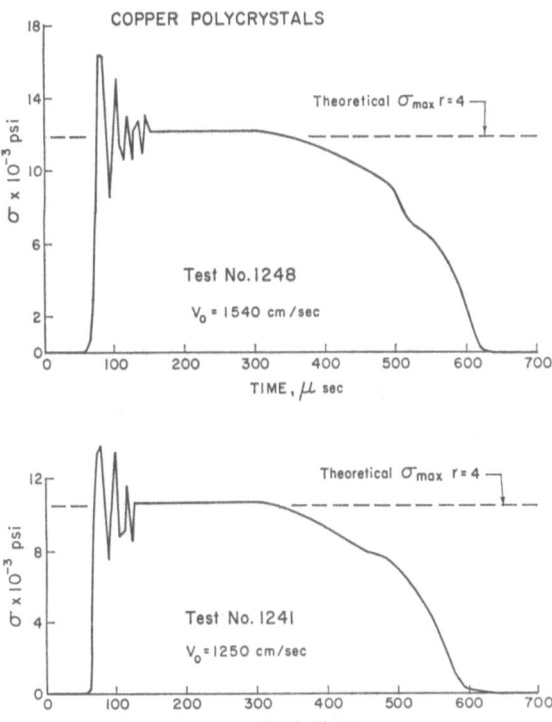

Fig. 2.38. Two load bar experiments in annealed polycrystalline copper,
compared with predicted maximum stress, Eq. (2.20), for $r = 4$

and with a 0.145 mm wall thickness (BELL and STEIN, 1962). Extensive
studies also have been made of wave propagation in specimens of different diameters to establish that irrespective of their magnitude the wave
initiation phenomenon was confined to the first diameter. In addition
to the 2.50 cm diameter solid cylinder studies described above, wave
propagation measurements have been made in specimens with 1.22 cm
diameter (HEINRICHS, 1961) and with 2.14 cm diameter (FILBEY,
1961a, b; BELL, 1967a). These data, together with the impact face
stress studies described above, provide ample experimental evidence
that the finite amplitude wave propagation behavior beyond the first
diameter occurs in the form of an uniaxial stress, plane wave front.

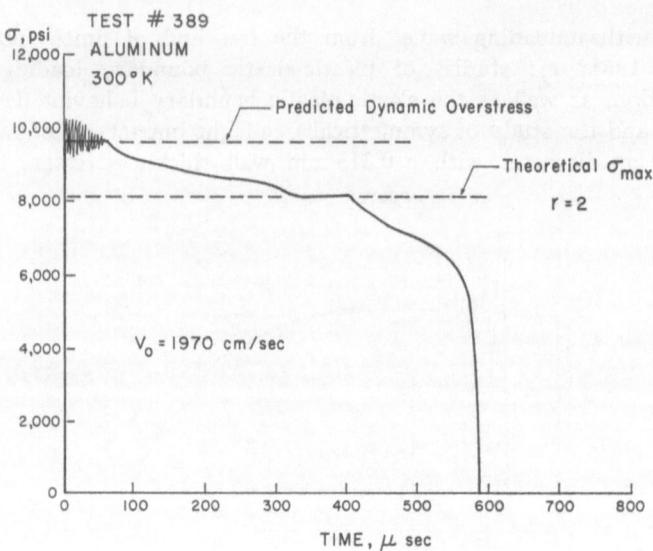

Fig. 2.39. A load bar experiment in polycrystalline aluminum, showing initial first diameter dynamic overstress with subsequent decrease to the theoretical stress of the parabolic stress-strain function, Eq. (2.20)

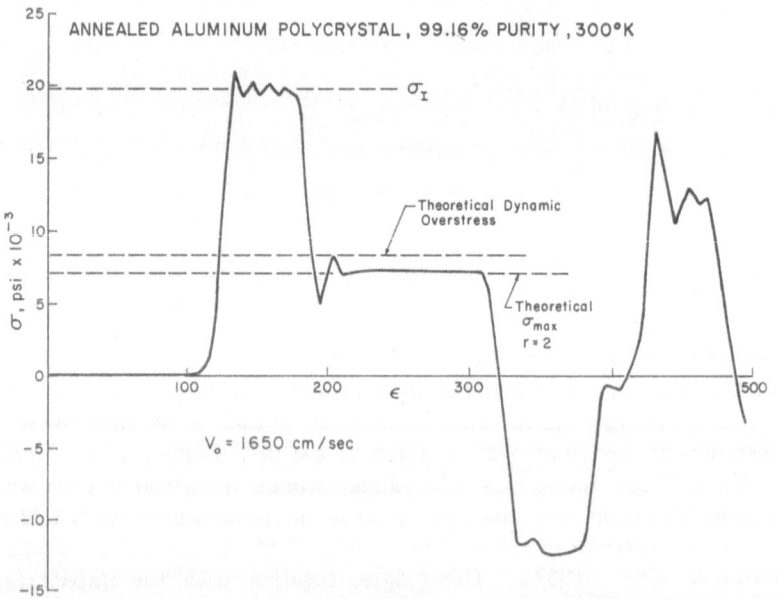

Fig. 2.40. Elastic-plastic boundary experiment in polycrystalline aluminum, showing rapid reduction to the theoretical stress of the parabolic law of Eq. (2.20)

Chapter III

Quasi-static Polycrystalline Uniaxial Stress Experiments

In the range of strain rates from 1 sec^{-1} to the highest diffraction grating measured value of 70,000 sec^{-1}, inertial contributions are sufficiently important so that the non-linear wave propagation effects of Chapter II must be considered. Strain rates of 1 sec^{-1} and below are in the domain of quasi-static, uniaxial stress experiments where this inertial contribution is negligible. For these experiments one need only establish that a uniform distribution of stress and finite strain is maintained with only the axial principal stress being non-zero. As will be shown below from the study of over 250 polycrystalline quasi-static, uniaxial stress experiments in nineteen crystalline elements and two binary combinations, the dynamic generalization of Eq. (2.20) is still applicable. As long as strain is increasing at the finite amplitude wave front in dynamic experiments, the integral value of the large deformation mode index r has been found to be stable until the maximum strain is reached. In quasi-static experiments, on the other hand, transitions from one parabolic large deformation mode to another are found to occur as the strain increases.

A central historical problem in the field of the large deformation of crystals has been the experimental question of whether or not viscosity plays a major role in finite distortional deformation behavior. In quasi-static experiments this question has been considered through the comparison of uniaxial stress-strain data from experiments in which either the deformation rate is maintained constant while the specimen length is varied, or deformation rates are varied for similar length cylindrical specimens. The most common test has been that of producing a nearly constant strain rate through the use of "hard" machines. The quantity to be determined is the stress-time history required to produce this prescribed deformation. Uniaxial stress may then be expressed as a function of strain and strain rate. In many polycrystalline metals, such experiments provide evidence that increasing the specified average strain rate results in an increase in the *stress* for a given value of the strain. Efforts to establish a functional relation between stress, strain, and strain rate from such observations have led to a fairly extensive collection of empirical proposals. These relations, fitted to quasi-static data have been extrapolated to high strain rates by numerous experimentalists who sought to separate the various proposals. That those extrapolations have led to contradiction and confusion in defining the role of viscosity in the finite deformation of crystalline solids is an historical fact. Such extended quasi-static impact experiments were plagued with extremes of non-reproducibility when data of one experimentalist were compared with another examining the same crystalline solid. In recent papers, the writer (BELL, 1965b, 1966b, 1967b) has described the historical logical difficulties of such experiments and, from direct strain measurement, has shown the limitations of the extended quasi-static impact experiments in providing any conclusion with reference to the role of viscosity in the large deformation of crystalline solids.

The observed difference in genuine quasi-static uniaxial stress experiments when low strain rates are varied is demonstrated here to be describable in terms of the present generalization, with respect to the stability properties of a finite deformation mode and transition structure. At sufficiently high strain rates, the large deformation stress-strain function is stable. This stability also characterizes quasi-static experiments in polycrystals of sufficiently high purity or at sufficiently low temperatures, and, as will be shown in the next chapter, it is also a property of the resolved deformation of high purity cubic single crystals.

In examining uniaxial quasi-static polycrystalline data with respect to the present parabolic large deformation generalization which includes transition phenomena, Eq. (2.20) is rewritten as Eq. (3.1),

$$\sigma = (2/3)^{r/2} \mu(0) B_0 (1 - T/T_m) (\varepsilon - \varepsilon_b)^{1/2} \tag{3.1}$$

where ε_b is a predictable intercept on the strain abscissa for the parabola or large deformation mode of interest.

For the uniaxial stress quasi-static experimental data, in terms of this generalization, three questions are paramount: a) Is the deformation parabolic? b) Is the temperature dependence linear? and c) Is the distortional behavior described by a specified integral value r and proportional to the measured $\mu(0)$ of the solid? In Eq. (3.1) the temperature of the test, T; the melting point of the polycrystal of interest, T_m; the zero-point isotropic elastic shear modulus, $\mu(0)$; and the dimensionless universal constant, $B_0 = 0.0280$ are all specified for a given experiment. Only the mode index r is unknown. By rewriting Eq. (3.1) as Eq. (3.2),

$$\sigma^2 = (2/3)^r \mu^2(0) B_0^2 (1 - T/T_m)^2 (\varepsilon - \varepsilon_b) \tag{3.2}$$

and plotting σ^2 vs ε, one may discern from the straightness of the experimental data whether the deformation is parabolic. In the event that parabolicity is observed, one may then determine whether or not the

observed linear slope corresponds to that for a given value of the integer
r ($r = 0, 1, 2, 3, 4, 5, ---$).

In σ^2 vs ε plots the difference in slope between parabolas with an
integral mode index r and with adjacent mode indices $r + 1$ or $r - 1$, is
2/3. Such differences are shown in Fig. 3.1.

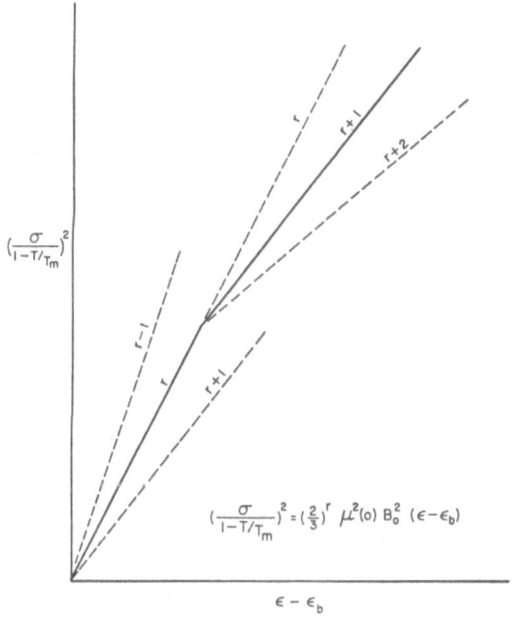

Fig. 3.1. A schematic σ^2 vs ε diagram of Eq. (3.2), showing magnitude of slope
differences for adjacent values of the mode index, r

Also schematically shown is the slope comparison for a parabola
with a deformation mode index r, undergoing a unit transition to a
second parabola with a deformation mode index of $r + 1$. The theoretical
slope variations of Fig. 3.1 are sufficiently large to permit a correlation
of experimental data with prediction. In all the data described below,
circles indicate experimental points and solid lines indicate theoretical
slopes (Eq. 3.1) for the designated value of the mode index r. The experimental point is at the center of the circle in every instance, and the size
of the circle is in no way to be construed as indicating possible variation
of the data.

A very large fraction of the quasi-static polycrystalline experimental
data in the literature is given in the form of logarithmic strain and
reduced stress, referred to in the metallurgical literature as "true"
stress and "true" strain. Since the generalized parabolic stress-strain
function, Eq. (3.1), obtained from dynamic studies refers to the *unde-*

formed state of material, the quasi-static, uniaxial stress data of the many authors cited below had to be "detrued" before comparing it with prediction. The choice of deformation measure is, of course, arbitrary but the parabolic generalization of Eq. (3.1) applies only when the data are expressed as *nominal* or engineering stress-strain functions.

Aluminum

The quasi-static, uniaxial stress large deformation experiment entails a relatively simple situation. It involves the measurement of only two quantities over a large range of values, both of the quantities being determined with a high degree of accuracy. Because of this simplicity, and the accuracy of measurement, any meaningful generalization should be capable of including the major portion of such experimental data.

A large segment of the macroscopic experimental data obtained during the past 40 years to support hypotheses introduced in various atomistic theories of finite crystalline deformation, have been for aluminum. In the literature the number of experimental studies on aluminum polycrystals and single crystals far exceeds those for any other crystalline element. Therefore, it is interesting to note that aluminum has a more complex mode and transition structure than most other metals of similar purity and metallurgical history.

The present writer's work is no exception to the emphasis upon the study of aluminum. In addition to a wide variety of finite wave propagation experiments totaling several hundred individual measurements (BELL, 1960b, 1961b, 1963a, 1967a), over seventy-five quasi-static polycrystalline experiments and a similar number of single crystal experiments have been performed. From the point of view of the deformation mode and transition structure, which is of major interest in the present monograph, the prevalence of instabilities in aluminum has been of importance in developing the large deformation generalization itself.

One of the most informative quasi-static, uniaxial stress experiments which has been carried out in the writer's laboratory is that for which the stress rate in a dead-weight loading test is maintained constant. The experimental apparatus for this test is shown in Fig. 3.2.

Through suitable controls the rate of water flow may be maintained constant. The finite strain is determined by a clip gauge over a predetermined length of specimen whose length to diameter ratio is approximately twelve to one. This experiment, modeled after earlier experiments of McREYNOLDS (1949), has been described in detail by SHARPE (1966a, b, c).

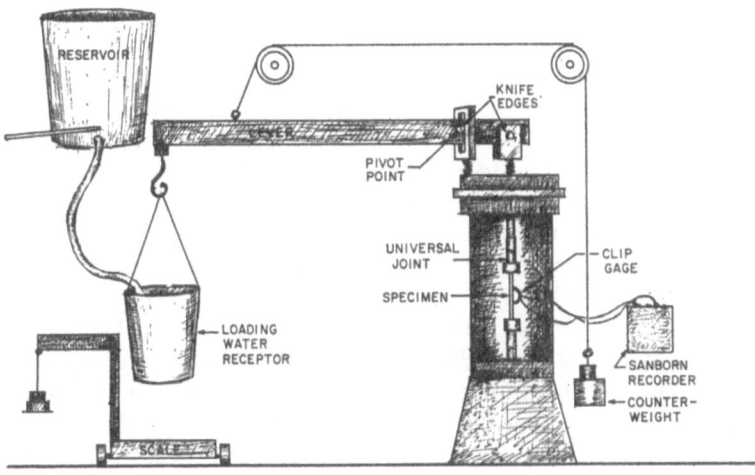

CONSTANT STRESS RATE APPARATUS

Fig. 3.2

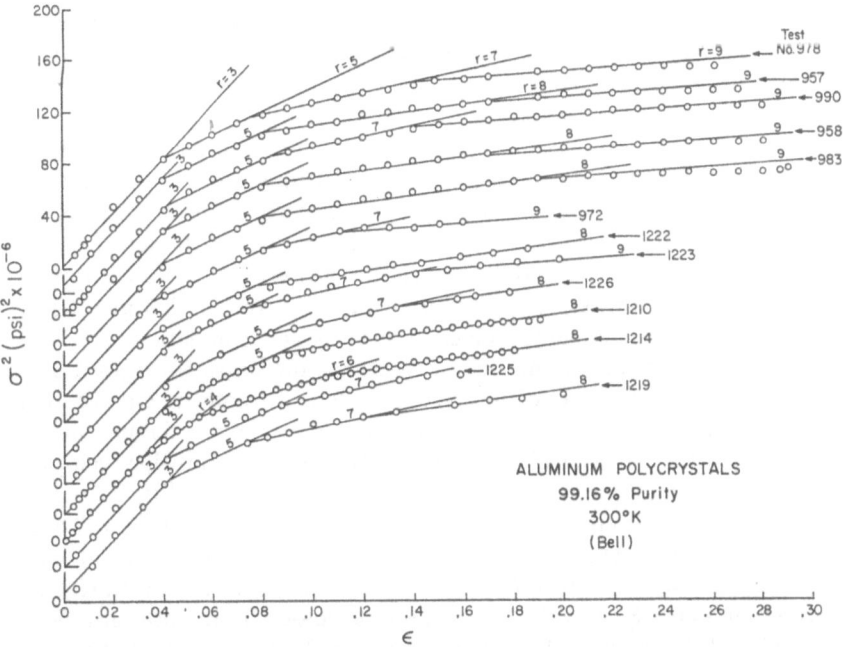

Fig. 3.3. 13 constant stress rate uniaxial stress tension experiments in low purity polycrystalline aluminum (circles) showing the reproducibility of deformation mode transitions at the second and third critical strain. The solid lines are theoretical prediction from Eq. (3.1)

Uniaxial stress-strain tension data for a constant stress rate of 5 psi sec^{-1} are shown in Fig. 3.3.

The specimens are 3/8 in. diameter; the gauge length is 2 in. The variable strain rate for this constant stress rate experiment is in the range of $\dot\varepsilon = 10^{-5}$ sec^{-1}. The material is 99.16% fine-grained polycrystalline aluminum, annealed for two hours at 867° K and furnace cooled. This annealing procedure was chosen so that the data could be compared with those from the finite amplitude wave studies considered in Chapter II.

All of the experiments in Fig. 3.3 were carried out in the writer's laboratory. The 900-series experiments are from the interesting study of the effects of grain boundaries on the Portevin-le Chatelier effect carried out by SHARPE (1966a, b, c). The 1200-series are experiments carried out for the present writer by R. KERSHNER, jr. These latter data, shown here for the first time, were obtained for the purpose of studying maximum straining rates for strain increments and elastic moduli changes associated with the rising portion of the Portevin-le Chatelier steps. Both these aspects of the Portevin-le Chatelier effect and their relation to the present large deformation mode and transition phenomenon will be considered in a later chapter. Of major interest here is that, in a σ^2 vs ε plot, the data fall into a series of reproducible straight lines corresponding to an initial theoretical slope of $r = 3$, with subsequent transitions to $r = 5, 7$, and 9 in nearly every instance.

As described in Chapter II, the dynamic studies of the writer for this same material, at strain rates from 5—9 orders of magnitude higher than $\dot\varepsilon = 10^{-5}$ sec^{-1}, have given for the rising strain front a stable parabola of $r = 2$ to strains as high as 17%. It was noted, however, that as the impact velocity was increased from the low values of the linear elastic range to high velocities, the wave initiation behavior revealed a series of transition velocities, calculated from the finite amplitude wave theory for the relevant dynamic polycrystalline large deformation parabola, $r = 2$. The critical strains at these transition velocities were 1.5%, 4.2%, and 7.5%. It is of interest, therefore, to note in Fig. 3.3 that the average value of the first transition from $r = 3$ to $r = 5$ reproducibly occurs at the same strain as that of the second transition velocity, $\varepsilon_N = 4.2\%$; and that the second quasi-static transition in Fig. 3.3 occurs at an average strain close to the third transition strain of $\varepsilon_N = 7.5\%$.

The dynamic experiments were carried out in compression at high strain rates, whereas the present experiments were carried out in tension at strain rates 8—9 orders of magnitude lower. That both situations provide the same fixed transition strains in the framework of the same generalized parabolicity is indeed remarkable.

To obtain the reproducibility of transition strains shown in Fig. 3.3, it is desirable that the crystalline solid be in a completely annealed state.

It is also preferable, but not essential, that the quasi-static, uniaxial stress experiment be performed on a dead-weight loading apparatus; i.e., the stress history is prescribed. In Chapter IV will be discussed single crystal studies in high purity aluminum where lower initial parabolas are stable through the first group of transitions. These data have provided an additional group of transitions corresponding to polycrystalline strains of 11.5%, 16.3%, and 26.6%. These transition strains are in addition to those observed at 1.5%, 4.2%, and 7.5%.

The 900-series experiments of Fig. 3.3 were carried to failure. It occurred in the vicinity of 27%, the sixth transition strain from single crystal prediction. The initial deformation of the thirteen tests of Fig. 3.3 proceed without a change of parabola coefficient through the first transition strain of 1.5% to the second value of 4.2%. This initial parabola of $r = 3$ provides a stress-strain function which lies below that obtained for the dynamic experiments ($r = 2$) at strain rates several orders of magnitude higher. Even though experimental conditions are carefully maintained constant, the initial parabola mode index may vary in quasi-static tests. Six tension experiments (786, 787, 788, 914, 1204, and 1207) of Fig. 3.4 are identical in every respect with the experiments of Fig. 3.3, including the constant stress rate of 5 psi sec^{-1}. The experiments of Fig. 3.4, however, begin with an $r = 2$ parabola as in the dynamic experiments at high strain rates, and show in each instance a transition to an $r = 3$ parabola in the vicinity of the first transition strain of 1.5%.

Experiment 971 in Fig. 3.4, which is also a constant stress rate tension measurement at 5 psi sec^{-1} was performed on 99.71% purity aluminum polycrystal. This experiment furnishes an initial parabola of $r = 3$ similar to the data of Fig. 3.3, but in this instance a transition occurs to $r = 5$ at the first transition strain of 1.5%, instead of at the second transition strain of 4.2%.

It is thus seen that the transitions, when present, occur at specified strains in annealed aluminum, irrespective of the parabola governing the deformation when it occurs. Uniaxial tension is accompanied by tensile normal stress components across the slip plane. Compression experiments, of course, provide a compressive normal stress across these slip planes. Therefore, it is entirely consistent that sharper transitions are found in the tensile situation than in the compressive. Thus, in Figs. 3.3 and 3.4 we see that for the tensile experiments the parabola coefficients proceeded in jumps involving two units of r, whereas the compressive tests exhibited consecutive transitions; i.e., in Fig. 3.3 tensile values of r are 3, 5, 7, and 9, whereas in Fig. 3.4 the compressive values of r are 2, 3, 4, and 5, as may be seen from an examination of the tests 1957-1, 2, 3. An exception to this behavior is the first transition for the tensile tests of Fig. 3.4 referred to above, which have an initial parabola of $r = 2$, followed by double transitions with a unit transition to $r = 3$ at the first transition strain. Most of the data shown in Fig. 3.4 are quasi-static uniaxial compression experiments. They are compared with the stress-strain parabola from 485 dynamic finite amplitude wave

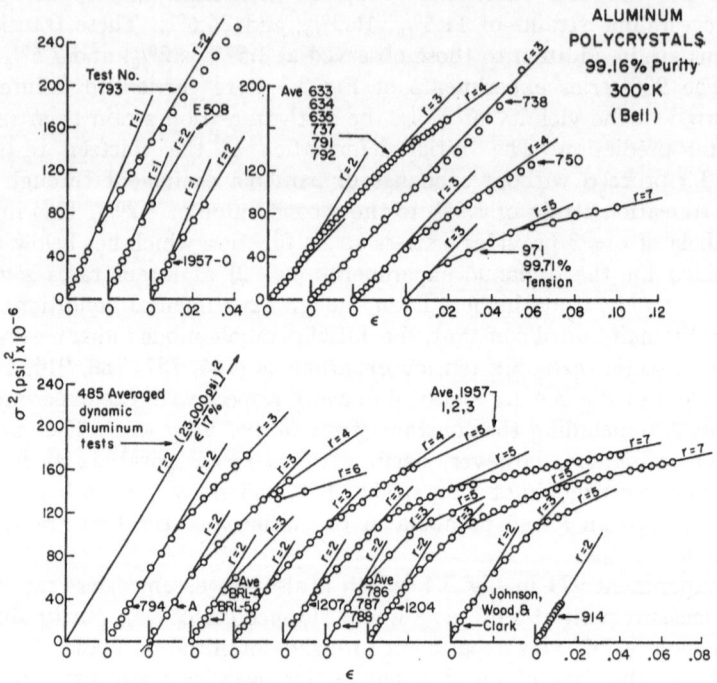

Fig. 3.4. Uniaxial stress experiments in low purity polycrystalline aluminum (circles) showing a variety of parabolic deformation modes and transitions. Experiments 786, 787, 788, 914, 971, 1204, and 1207 are in tension. All other experiments are in compression. The solid lines are theoretical prediction from Eq. (3.1)

propagation experiments in which the $r = 2$ parabola has been observed by means of diffraction gratings to maximum strains as high as 17%.

Several years ago the writer pointed out that quasi-static uniaxial compression experiments in 99.16% purity polycrystalline aluminum provided the same stress-strain curve as that which governed the wave propagation ($r = 2$) to strains of approximately 2% (BELL, 1960b, 1961b). While in general this is the predominant quasi-static compression behavior, some exceptions have been observed. These are tests E 508, 1957-0, and 750 of Fig. 3.4 for 99.16% aluminum annealed for two hours and furnace cooled in the same manner as all of the other aluminum polycrystalline specimens described thus far. The first two experiments provided an initial parabola of $r = 1$ with a transition to $r = 2$ at approximately the first transition strain of 1.5%, and the third experiment provided an $r = 3$ parabola with a transition to $r = 4$ at a strain somewhat larger than the first transition strain of 1.5%. As in other compression measurements, these transitions involve a unit change in the value of r.

Almost all of these uniaxial compression experiments have been performed with lubricated specimen interfaces. Two exceptions are E 508 and 1957-0, which were experiments performed twelve and ten years ago respectively, and for which the lubrication situation is uncertain. Experiment 793 is a lubricated compression experiment which also has an $r = 1$ parabola with a transition to $r = 2$ at the first transition strain of 1.5%. This specimen differed from the others in its prior metallurgical history although it was annealed in the same manner in the writer's laboratory. The material for test 793 has the designation of 1100-H-18.

A most interesting series of compression experiments in Fig. 3.4 are designated as 633, 634, 635, 737, 791, 792, and 738. These experiments were performed to produce differences in the transition behavior and, in particular, in the mode index of the initial parabola. Test 738 is for a conventional 99.16% purity fine-grained aluminum polycrystal which initially produced an $r = 2$ parabola. The specimen was then melted and repoured as a large-grained polycrystal. A lubricated uniaxial compression measurement was then made which provided the $r = 3$ parabola to a strain of over 12% (see Fig. 3.4). A similar experiment, 737, provided the curious behavior shown in Fig. 3.4. The initial parabola is $r = 3$, as test 738, but a first transition occurred at the strain of 1.5% with a reduction in the value of r from 3—2, immediately followed by a second transition back to $r = 3$. At a much larger strain a unit transition occurred with r increasing from $r = 3$ to $r = 4$. Experiments 633, 634, 635, 791, and 792 have an identical behavior with that of test 737, but in these experiments the specimens are the original annealed 99.16% purity polycrystals. These experiments differ from the standard compression measurements in that the specimen end faces were in each instance encased in teflon, which allowed a relatively unrestrained grain rotation at both specimen interfaces. The quasistatic uniaxial stress test of JOHNSON, WOOD, and CLARK (1953) has been included for comparison with the writer's measurements in Fig. 3.4. This test is from a 1953 paper which considered the dynamic overstress produced in the Hopkinson load bar experiment (BELL, 1960c, 1963b).

From the thirty-eight 99.16% purity quasi-static uniaxial stress experiments of Figs. 3.3 and 3.4, one may see that for this low purity

completely annealed polycrystalline solid the large deformation behavior
is in every instance parabolic; moreover, the experimental parabola
coefficients agree with theoretical slopes for specified values of r (solid
lines) when compared with the experimental data (circles). Reproducibility is observed for the transition structure in the quasi-static experiments as well as for the second-order transitions which occur at a specified set of strains. As will be shown in Chapter IV, the same transition
structure is observed when studying the resolved shear stress, resolved
shear strain, large deformation of 99.16% purity single crystals.

All of the compression experiments of Fig. 3.4 and Figs. 3.5 and 3.6
were uniaxial stress experiments on one-inch diameter, 3 in. long,
lubricated specimens. All of the tension tests of Figs. 3.3, 3.4, and 3.5
were for 3/8 in. diameter specimens with a 4 in. long test section. In
Figs. 3.3 and 3.4, all of the data except test 971 were for 99.16% purity
specimens. A similar series of tension and compression uniaxial stress
experiments were performed in 99.45% and 99.99% purity polycrystals.
These data are shown in Fig. 3.5.

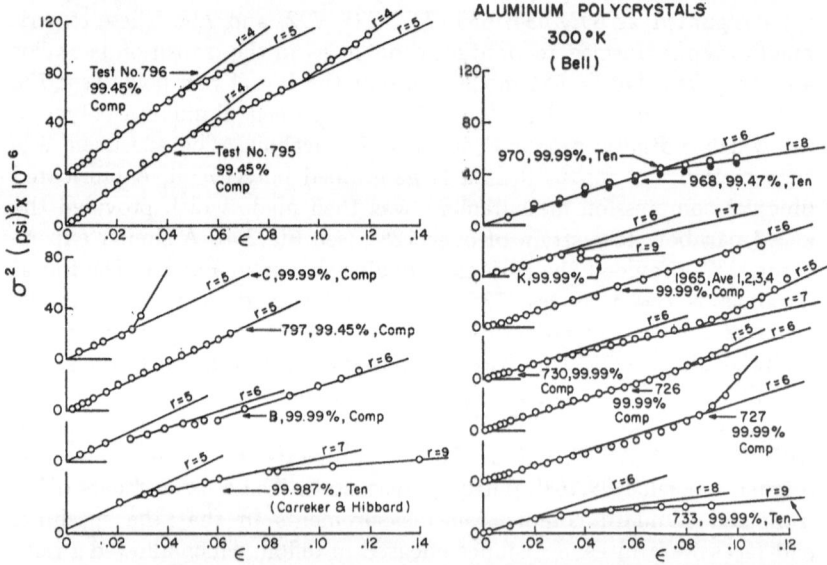

Fig. 3.5. Tension and compression experiments in polycrystalline aluminum
of medium to high purity (circles) compared with prediction from Eq. (3.1)
(solid lines) for the indicated mode indices, r

As in Figs. 3.3 and 3.4 the experimental data (circles) are compared
with the theoretical slopes (straight lines) in σ^2 vs ε plots. The 99.45%
aluminum polycrystals in compression are found to be parabolic with

initial parabolas in two instances (tests 795 and 796) with a deformation mode index $r = 4$ to the second transition strain of 4.2%. At 4.2% strain unit transitions to $r = 5$ occur in each experiment, as is characteristic of compression measurements. One experiment at this purity, 797, had an initial parabola of $r = 5$ which continued until the test ended at not quite 7% strain. In the 99.47% purity aluminum polycrystal uniaxial stress tension experiment, test 968 has an initial parabola of $r = 6$ to the vicinity of the third transition strain, $\varepsilon = 7.5\%$. At this strain a double transition occurs to $r = 8$, as is characteristic of tension data. This test 968 and the 99.99% purity tensile test, 970, were performed in the constant stress rate apparatus, Fig. 3.2, at the same constant stress rate (5 psi sec^{-1}) as the experiments described in Fig. 3.3. It is interesting that this 99.99% purity test is identical in initial parabola and transition with the 99.47% purity tension test. Both of these experiments were performed by SHARPE (1966a, b, c) in his study of the influence of grain boundaries upon the proliferation of the serrations of the stress-strain function known as the Portevin-le Chatelier effect. Despite the fact that the phenomenon disappeared with the increase in purity in these two experiments of SHARPE, the mean stress-strain function of experiments 970 and 968 are identical.

The 99.99% purity compression tests K, 1965-1, 2, 3, 4, 726, 727, and 730 all have initial parabolas of $r = 6$. Tests 1965-1, 2, 3, 4 reached 11% without a transition. At this strain the specimens buckled and the experiment ceased. It is interesting that single crystal tests, as was mentioned above, predict a transition at 11.5%. The uniaxial compression experiment, test K, is one described by the writer in an earlier paper (BELL, 1963b), where it was shown that for this high purity solid the dynamic non-linear wave propagation uniaxial stress-strain function is identical with the high purity quasi-static function ($r = 6$), despite the fact that strain rates varied by seven orders of magnitude.

All of the 99.99% purity specimens were prepared by the writer from high purity ingots. The specimens were poured into cylindrical molds with the furnace temperature being reduced to room temperature under controlled conditions over a period of time. This method of specimen preparation produced annealed polycrystals with large grain size. A series of x-ray diffraction measurements were made on the individual grains of such a polycrystalline specimen to insure that the initial crystallographic orientations were sufficiently random. These pre-deformation measurements, together with x-ray diffraction observations of the same grains after deformation, have shown that this is indeed the situation. This experiment 921 is described in detail in the next chapter.

Two tests in the series are of particular interest; i.e., tests B and C. In an effort to produce a change in the parabola coefficient r, the furnace in which these specimens were poured was mounted on a vibrating table both during the pouring of the specimen and during the subsequent cooling. As is characteristic of the other high purity columnar multi-crystal compression specimens, each grain has one free surface. For the vibrated specimens, however, these grains have a much smaller diameter needle shape. Both tests B and C have an initial parabola mode index of $r = 5$ or a unit difference from the standard compression multicrystal value of $r = 6$. Test C underwent an abrupt upward transition to $r = 2$ at a strain slightly above 2%, and test B underwent a unit transition from $r = 5$ to $r = 6$ at the first transition strain, $\varepsilon_N = 1.5\%$. A second and very interesting shift in the deformation occurred at the second transition strain, $\varepsilon_N = 4.2\%$, but the parabola retained the $r = 6$ coefficient.

The transitions described in this monograph are all second order; i.e., a discontinuity occurs in the derivative of the stress-strain function. The vibrated compression uniaxial stress experiment B is unique in that it has a transition which would appear to be of the first order. The maximum compression strain at which specimen B buckled is close to the predicted fourth polycrystalline transition strain of 11.5%. As will be shown in Chapter IV, this polycrystalline transition strain is predictable from single crystal experiments.

The 99.99% purity, uniaxial tension experiment of test 733 (Fig. 3.5) has an initial parabola of $r = 6$, then undergoes a double transition to $r = 8$ at the first transition strain, and a unit transition to $r = 9$ at the second transition strain. CARREKER and HIBBARD's (1957) 99.987% purity large-grain tensile test at $300°$ K, for which double transitions occur from $r = 5$ to $r = 7$, and from $r = 7$ to $r = 9$, is included in Fig. 3.5 for comparison with the present writer's room temperature data. GILLICH (1964) has shown that this experiment of CARREKER and HIBBARD has a fourth power deformation envelope which may characterize transition structure in different metals for fractional melting point temperature in the vicinity of $T/T_m = 1/3$. Other power laws may be shown to be deformation envelopes at different values of T/T_m. The compression series 1957-1, 2, 3, of Fig. 3.4 were shown by the writer (BELL, 1963b) to have a 3/8 power law envelope. The significance of empirically describing these data in terms of their envelopes is still to be determined.

A comparison of the low purity data of Figs. 3.3 and 3.4 with the higher purity data of Fig. 3.5 reveals that for both tension and compression uniaxial stress experiments, the generalized parabolic stress-strain function, Eq. (3.1), is applicable. The effect of increasing the purity is to increase the value of r for the initial parabola and to suppress the transitions at lower values of strain.

All of the data of Figs. 3.3, 3.4, and 3.5 are uniaxial stress experiments at room temperature; i.e., 300° K. Figs. 3.6 and 3.7 show a series of very low strain rate, high and low purity , dead-weight compression specimens which are 1 in. diameter, 3 in. long with lubricated interfaces. Two of the specimens of Fig. 3.6 are 99.16% purity, studied at a temperature of 423° K.

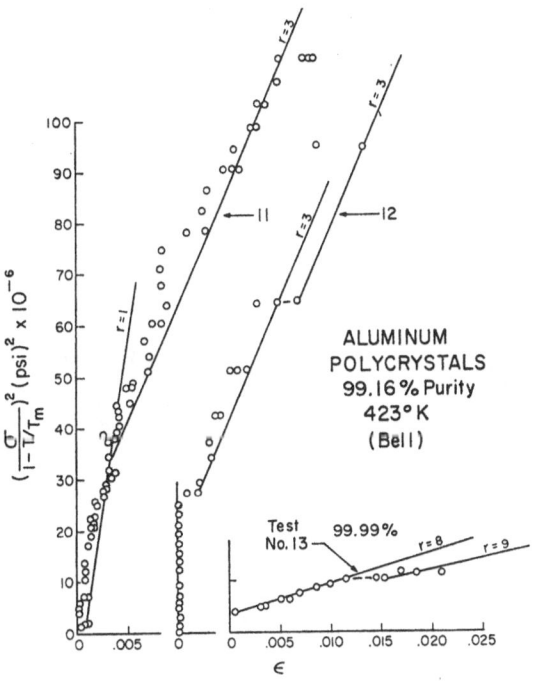

Fig. 3.6. Low purity uniaxial compression experiments in polycrystalline aluminum at 423° K (circles) compared with prediction from Eq. (3.1) (solid lines) for the indicated mode indices, r. Note the presence of the Portevin - le Chatelier effect

These low purity compression experiments at this higher temperature are dominated by the Portevin - le Chatelier steps, contrary to earlier (McReynolds, 1949) evidence in tension that the Portevin-le Chatelier effect disappeared with increasing temperature in low purity aluminum. This aspect of these experiments will be considered in Chapter VII. The important point here is that with an increase in temperature the initial parabola for the low purity solid has changed from $r = 2$ to $r = 3$.

For the high purity compression tests of Figs. 3.6 and 3.7, one notes the same behavior. Initial parabolas for these 99.99% purity aluminum polycrystals are either $r = 7$ or $r = 8$, instead of the dominant values of $r = 6$ at 300° K. In addition to the change in temperature, these experiments differ from the measurements in Fig. 3.5 in having been performed in a dead-weight compression apparatus at far lower strain rates. Thus, we see that the test at 293° K also has an initial parabola of $r = 8$. It is interesting that transition strains in these high temperature tests approximate the room temperature values described above. The effects of temperature upon the deformation mode and transition

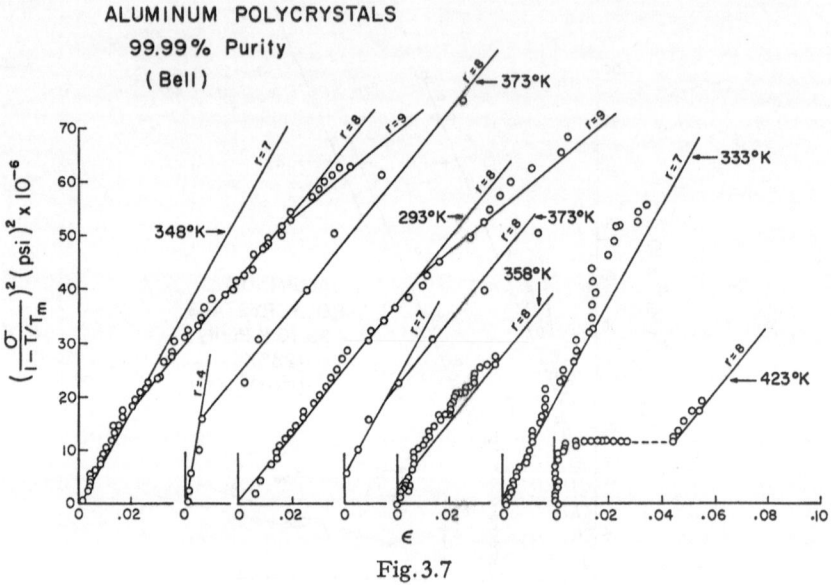

Fig. 3.7

structure can better be seen by comparing data over a wide range of temperatures. Such comparisons are shown in Fig. 3.8 for CARREKER and HIBBARD's (1957) high purity aluminum uniaxial stress tension experiments.

As in dynamic experiments, at 20° K a stable $r = 2$ parabola is observed to a strain of nearly 20%. One may note that $r = 2$ also is the deformation mode index for the dynamic large deformation of the 99.16% aluminum in the compression wave propagation experiments of the writer described in Chapter II. As the temperature increases, the value of the initial parabola deformation mode index r increases, and transitions appear. Thus at 77° K the initial deformation mode index is $r = 3$; at 140° K, $r = 5$; at 195° K and 300° K, $r = 7$; and at 398° K, $r = 8$. A comparison of the mode indices, r, at 300° K and 398° K in these high purity tension tests indicates that the values are similar to those in the writer's high purity compression experiments of Figs. 3.6

and 3.7. Despite the fact that the writer's compression experiments are dead-weight tests, whereas the experiments of CARREKER and HIBBARD (1957) are not, one still finds that the transitions occur at strains approximating the specified values determined from the dynamic compression measurements. It is of interest, too, that these tension uni-axial stress transitions exceeded a unit change in r, as was characteristic of the writer's tension experiments described above.

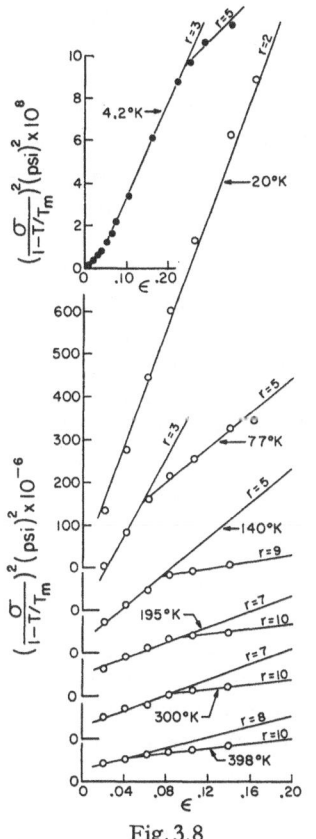

ALUMINUM POLYCRYSTALS
o – 99.987% Purity (Carreker & Hibbard)
• – 99.99% Purity (Hosford, et al.)

Fig. 3.8

Also shown in Fig. 3.8 is a uni-axial tension experiment of HOS-FORD, FLEISCHER, and BACKOFEN (1960) at 4.2° K for a 99.99% purity aluminum polycrystal. The initial low strain behavior in this experiment arises from the presence of a stage II polycrystalline deformation. This aspect of the experiment will be discussed in Chapter IV. One notes that the initial parabola at 4.2° K has a value of $r = 3$ in comparison with the value of $r = 2$ for CARREKER and HIBBARD's (1957) test at 20° K, indicating the same probability aspect of the initial deformation mode observed at room temperatures. This value for $r = 3$ of the HOSFORD et al. (1960) experiment remained stable to 24% when, as is characteristic of tension experiments, a double transition to $r = 5$ occurred.

As is characteristic of the data of all of the other elements to be considered in this chapter, the data of Fig. 3.8 are reduced to absolute zero by plotting $\left(\dfrac{\sigma}{1 - T/T_m}\right)^2$ against ε, thus allowing a somewhat clearer comparison of the behavior of the various crystalline solids. [As before, the circles are the experimental data and the solid lines are theoretical slopes in terms of Eq. (3.1) for the specified value of r.] It is important to emphasize that the data of both HOSFORD, et al. (1960), and CARREKER and HIBBARD (1957) were given in their respective publications in terms of reduced stress and logarithmic strain; i.e., "true" stress versus "true" strain. In both papers the method of this reduction of the original data was stated so that it was possible for the present writer to recalculate the data back to its original nominal form for comparison with the parabolic large deformation generalization of Eq. (3.1).

An inspection of the quasi-static uniaxial stress experiments in aluminum described in the present chapter, and the wave propagation uniaxial stress experiments in aluminum described in Chapter II, reveals that irrespective of purity, strain rate, specimen geometry, or whether the deformation is tensile or compressive, the distortional large deformation of aluminum is governed by Eq. (3.1).

That this linearly temperature dependent parabolic large deformation behavior has its origin in the shear deformation of the crystals which comprise the aggregate will be shown in Chapter IV. Before attempting to do anything other than describe this newly discovered large deformation parabolic mode and transition structure, it is of importance to determine its degree of generality among the elements and binary combinations. The remainder of the present chapter records the results of an extensive review of uniaxial stress experiments in the polycrystalline large deformation literature for nineteen elements and two binary combinations: Al, Ag, Au, Ni, Cu, Pb, Ir, Rh, Fe, Cr, Mo, Ta, Nb, Mg, Zn, Re, Y, Ge, U, 70—30 Cu-Zn, Mo-50 Re.

In the literature nearly all the experimental uniaxial stress data are presented in the form of "true" stress and natural strain. These data had to be "detrued" to their original nominal form by the present writer for comparison with the parabolic generalization, Eq. (3.1). The original data were first obtained in nominal form and then calculated as "true" stress-strain functions before being plotted; these same plots are now remeasured and recalculated by the present writer back to the original nominal form. Because of these multiple calculations and data plotting, the location of transition strain is subject to some variation; for this reason, no effort has been made to systematically use these data from the literature for the study of the magnitude of critical strains. Occasional reference will be made to the transition strain values of individual experiments. The close agreement of critical strains in α-brass, Cu, and Fe compared with Al will be shown below. It should be noted that most of the transition strains of the Ag, Au, and Cu experiments shown below do occur at the specified transition values, and that the actual critical strain values are *independent* of the temperature of the experiment.

Uranium

In examining different crystalline solids, one of the most interesting immediate questions is whether or not the parabolic behavior observed in aluminum is associated with its face-centered cubic structure. HOCKETT (1959) performed a series of uniaxial stress compression experiments on uranium which has an orthorhombic crystal structure. These experiments of HOCKETT were carried out at different constant strain rates for the purpose of studying the effect of strain rate upon the deformation. A cam plastometer was designed to compress lubricated 0.800 in. diameter specimens to 50% of their original height. The cam was also designed to produce a constant logarithmic strain rate. The solid was described as "relatively pure, depleted uranium". As is customary in strain rate studies, the data were presented in terms of reduced stress and logarithmic strain.

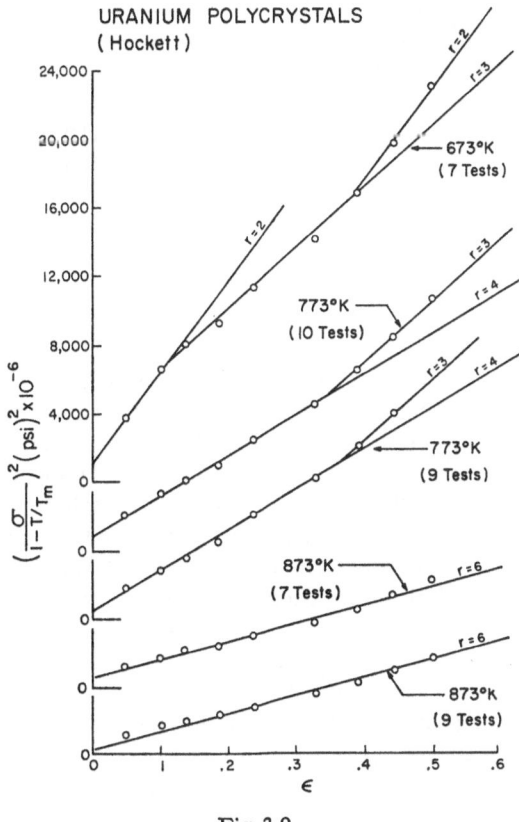

Fig. 3.9

The experimental data in Fig. 3.9 also have been "detrued" by the present writer to their original nominal form for a comparison with the writer's large deformation generalization. The original data are presented as averages of multiple tests for "true" strain rates — i.e., 10^{-1} sec^{-1} and 1 sec^{-1} — at three test temperatures of 673° K, 773° K, and 873° K. At all three temperatures these experimental data (circles), reduced to absolute zero by plotting $\left(\dfrac{\sigma}{1 - T/T_m}\right)^2$ vs ε are compared with theoretical straight lines (solid lines). It is found that these uranium data are in close agreement with the present writer's parabolic generalization when the data are shown as nominal (engineering) stress-strain curves. In each instance, of course, the maximum strain for these constant strain rate tests is 50%. As in the aluminum, the value of r for the initial parabola increases with increasing temperature.

It is interesting that all of the high purity tests at 873° K are stable on an $r = 6$ parabola to 50%. The data at both lower temperatures exhibit the same transition structure which characterizes aluminum. At 873° K and 773° K the lower curve represents in each instance experiments with strain rates of $\dot{\varepsilon} = 10^{-1}$ sec^{-1}, whereas the upper curve represents experiments with strain rates one order of magnitude higher; i.e., $\dot{\varepsilon} = 1$ sec^{-1}. This latter strain rate corresponds to a strain rate of quasi-static testing, above which inertial effects begin to be important. One sees that at both temperatures this increase of strain rate does not affect the parabola coefficients throughout the deformation but does change the intercept of the parabola on the strain abscissa. This has the effect of shifting upward the entire plastic portion of the stress-strain curve.

The reproducibility of the transition strains in the two sets of averaged data at 773° K shows that the transition strain from $r = 4$ to $r = 3$ also is shifted in a manner consistent with this change in the parabola intercept on the strain abscissa. As far as the large deformation generalization is concerned, these experimental data of HOCKETT (1959) show that crystallographically orthorhombic polycrystalline uranium is in close agreement with the present writer's large deformation generalization. The differences between the tests of Fig. 3.9 calculated at absolute zero, and the similarly calculated aluminum data of CARREKER and HIBBARD (1957) and of HOSFORD, et al. (1960), in Fig. 3.8, is solely that of comparing the relative values of the zero-point isotropic shear moduli, $\mu(0)$. As will be shown in Chapter V, uranium has an experimental value of $\mu(0) = 8,450$ kg/mm^2, whereas aluminum has a $\mu(0) = 3,110$ kg/mm^2.

It is also interesting that since Eq. (3.1) is written in terms of the fractional temperature ratio, T/T_m, the differences in the melting points of aluminum and uranium (932° K and 1,405° K, respectively) are such that the range of experiments are considerably different. For aluminum at the highest temperature tested the value of $T/T_m = 0.45$ at $423°$ K, Fig. 3.6, Fig. 3.7; whereas for uranium at $873°$ K, the tests have a higher value of $T/T_m = 0.63$.

Zinc

In Chapter II, 99.2% polycrystalline zinc was shown experimentally to have a finite amplitude wave propagation behavior consistent with that of the other polycrystalline metals studied; i.e., it was found that the finite amplitude wave theory was applicable. The governing stress-strain law was shown to be parabolic with a deformation mode index of $r = 2$. It was also shown that the strain-time detail could be predicted

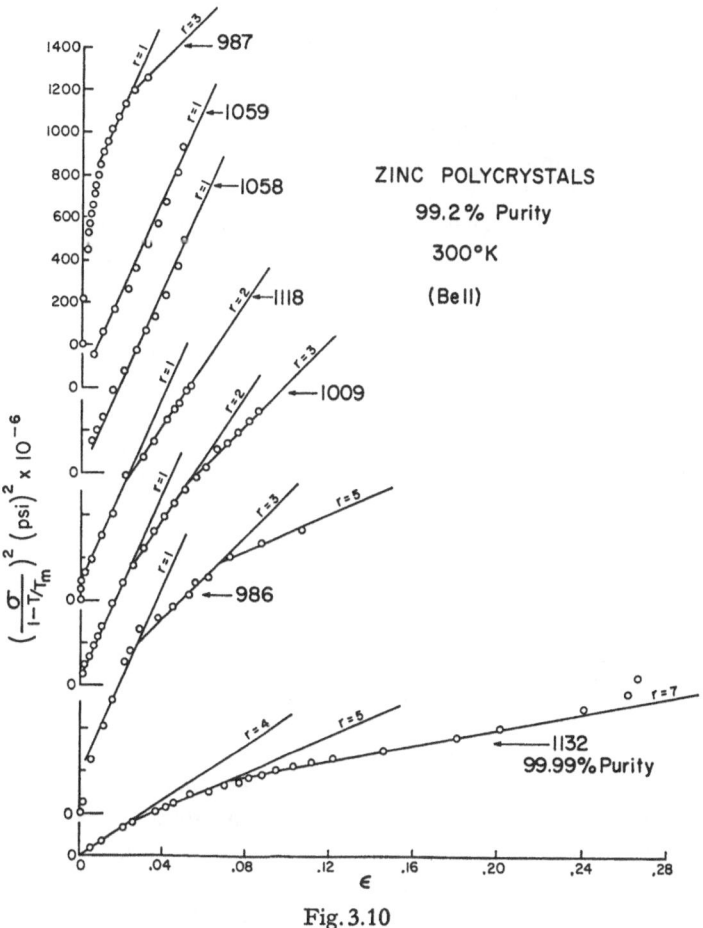

Fig. 3.10

from aluminum polycrystals of the same purity by merely noting the differences in mass density, melting point, and zero-point isotropic shear moduli. This is of particular interest because zinc, like orthorhombic uranium whose quasi-static data has been shown to fit the same generalization, Eq. (3.1), also differs in crystal structure. Zinc has an hexagonal crystal structure.

In Fig. 3.10 are shown six room temperature quasi-static uniaxial compression experiments in zinc for 99.2% purity. These specimens were annealed at 588° K for $2^1/_2$ hours and furnace cooled. An inspection of the $\left(\dfrac{\sigma}{1-T/T_m}\right)^2$ vs ε plot of Fig. 3.10 reveals that these experimental data (circles) not only fit predicted parabolas (solid lines), but also exhibit the large deformation mode and transition structure that is characterized by face-centered cubic aluminum.

In view of the common assumption among many experimentalists that a difference between quasi-static and dynamic stress-strain curves offers experimental evidence of the importance of viscosity in solids, these low purity zinc polycrystalline experiments provide them with a dilemma. The initial parabola for every low purity polycrystalline experiment has a deformation mode index of $r = 1$ for the quasi-static tests, whereas for the dynamic data the corresponding governing parabolic stress-strain curve has a deformation mode index of $r = 2$; i.e., the dynamic stress-strain function at high strain rates lies below the quasi-static stress-strain function obtained at strain rates several orders of magnitude lower. As in aluminum, the increase in purity has the effect of increasing the numerical value of the deformation mode index. As may be seen from Fig. 3.10, all the zinc experiments were performed at 300° K. The data shown have been reduced to absolute zero for purposes of comparison. The increase in purity changes the initial deformation mode index from $r = 1$ to $r = 4$.

Germanium

Polycrystalline germanium, for which no dynamic wave propagation data is available, represents still another crystal type; i.e., germanium has a diamond crystal structure. Nevertheless, as may be seen in Fig. 3.11, which shows a σ^2 vs ε plot of an uniaxial compression test of PATEL and ALEXANDER (1956) at 873° K, the large deformation is parabolic having an initial parabola with a deformation mode index of $r = 2$. The finite deformation undergoes an upward transition to $r = 0$ at a strain of 1.5%. For this element insufficient data is available to know whether there is significance in the fact that this transition occurs at 1.5% strain, which is the first transition strain of aluminum.

Again, as for uranium and zinc, the comparison of the experimental data (circles) with the predicted parabola (solid lines) is based upon the ratio of the differences in melting points, T_m, and zero-point isotropic shear moduli, $\mu(0)$. These comparisons between aluminum and germanium are 932° K and 1,231° K, 3,110 kg/mm² and 7070 kg/mm², respectively. The T/T_m value of the Ge test of Fig. 3.11 is much higher than for any of the quasi-static aluminum data; i.e., $T/T_m = 0.71$. Nevertheless, at this high fractional

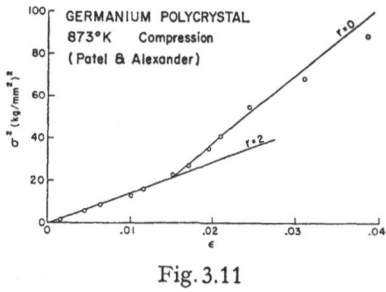

Fig. 3.11

melting point temperature, despite the difference in crystal type, the uniaxial stress experiment for germanium has a deformation mode index that is parabolic and is consistent with the mode and deformation structure of Eq. (3.1).

Tantalum

A quasi-static uniaxial stress experiment in still another crystal type is shown in Fig. 3.12. Tantalum has a body-centered cubic crystalline structure. It has a melting point of 3,269° K which is far above the value of 932° K for aluminum. Its zero-point isotropic shear modulus of 7,040 kg/mm² also is far above the 3,110 kg/mm² for aluminum. Nevertheless, this uniaxial tension experiment of BECHTOLD, WESSEL, and FRANCE (1961) at 573° K in tantalum fits the generalization of Eq. (3.1). The initial parabola has a deformation mode index of $r = 0$; it undergoes

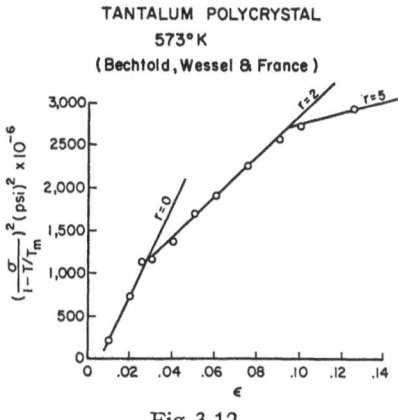

Fig. 3.12

a double transition to $r = 2$ at approximately 2.5% strain, and subsequently to $r = 5$, as is characteristic of tension experiments in face-centered cubic aluminum. BECHTOLD *et al.* (1961) lower temperature tests in tantalum not only did not exhibit parabolic large deformation behavior, but did not show an increase in stress in the plastic region after the elastic limit had been reached. However, a room temperature test of JAFFEE, MAYKUTH, and DOUGLASS (1961) shown in Fig. 3.13 does provide an agreement with the present writer's parabolic generalization at 300° K.

POLYCRYSTAL DATA OF JAFFEE, MAYKUTH & DOUGLASS
300°K

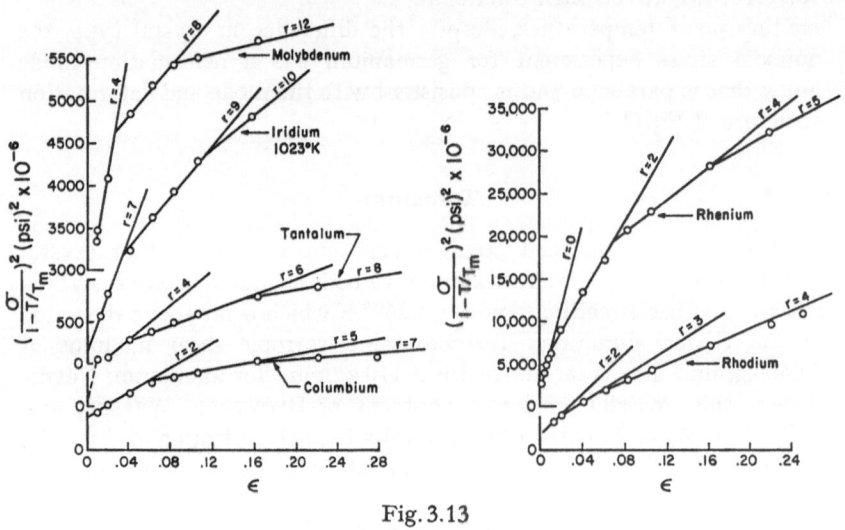

Fig. 3.13

Iridium, Rhenium, Molybdenum, Rhodium, and Niobium

The uniaxial tension tests for the five elements of Fig. 3.13 had to be "detrued" to the original nominal form for comparison with the parabolic generalization. In each instance, the experimental data (circles) and the predicted parabolas (solid lines) are found to provide an agreement with the generalization. Molybdenum, tantalum, and niobium (columbium) have a body-centered crystalline structure, rhodium a face-centered crystal structure, and rhenium a hexagonal crystal structure. A wide range of melting points T_m and zero-point isotropic shear moduli $\mu(0)$ is found when these elements are compared with aluminum or zinc.

For molybdenum $T_m = 2,830°$ K and $\mu(0) = 15,400$; for iridium $T_m = 2,728°$ K and $\mu(0) = 23,100$. Iridium and osmium have the highest zero-point isotropic shear moduli of any of the fifty-eight elements and five binary combinations considered in this monograph. As pointed out before, tantalum has a melting point of 3,269° K and $\mu(0)$ of 7,040. Niobium (columbium) has a T_m of 2,743 °K and a $\mu(0)$ of 3,900. Rhenium has the high melting point, T_m of 3,443° K and also a high $\mu(0)$ of 21,400. Rhodium has a T_m of 2,239° K and a $\mu(0)$ of 15,830 kg/mm². Despite the fact that these elements have melting points T_m and zero-point isotropic shear moduli $\mu(0)$ far above those for the other elements considered in this monograph, their uniaxial tension stress-strain functions provide a remarkably close agreement with the writer's generalized parabolic large deformation behavior of Eq. (3.1). All of these data are reduced to absolute zero for purposes of comparison by plotting $\left(\dfrac{\sigma}{1 - T/T_m} \right)^2$ vs ε. One notes that for each of these crystal structures the finite deformation mode and transition structure originally discussed by the writer in aluminum (BELL, 1961b, 1963b, 1967a; BELL and SUCKLING, 1962) is found to occur.

Gold

An examination of the uniaxial stress quasi-static experiments presented thus far reveals that the writer's linearly temperature dependent, parabolic large deformation generalization is still applicable even when one varies the crystallographic type, purity, ambient temperature, specimen diameter, specimen length, strain rate, and type of uniaxial loading; i.e., tension or compression.

The 99.99% purity polycrystalline gold experiments of McKEOWN and HUDSON (1937) permit of the examination of still another variable, the prior work-hardening of the solid. Gold, which is face-centered cubic in crystal structure, is of particular interest in the present study because its zero-point isotropic linear elastic shear modulus, like that of silver, is nearly identical with aluminum. (For gold, silver, and aluminum values of $\mu(0)$ are 3,090 kg/mm², 3170 kg/mm², and 3110 kg/mm², respectively.) The melting point of gold is considerably higher than that of aluminum; i.e., $T_m = 1,334°$ K, compared to 932° K. During the period in which the present large deformation generalization was being developed, observations of the similarity of parabola coefficients in aluminum, silver, and gold were made. The effort to understand why these three elements should have common parabola coefficients, despite the differences in their melting points and densities, was an important factor in the discovery of the proportionality of these coefficients to the zero-point shear moduli.

The two uniaxial stress tension experiments of McKeown and Hudson (1937) in gold, shown in Fig. 3.14, are for 0.564 in. specimens annealed at 773° K.

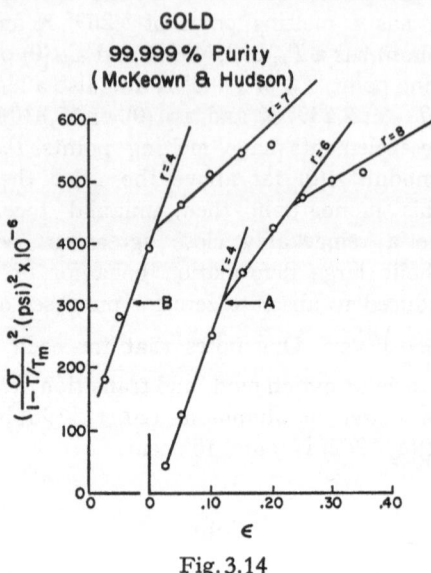

Fig. 3.14

The test designated as "A" in the $\left(\dfrac{\sigma}{1 - T/T_m}\right)^2$ vs ε plot of Fig. 3.14 is for an annealed specimen. It has the characteristic tensile double transitions. The deformation obviously is parabolic. The first transition from $r = 4$ to $r = 6$ occurs at the fourth transition strain of 11.5%. The second transition from $r = 6$ to $r = 8$ occurs at a value of strain in the vicinity of the sixth transition strain of 27%.

Test "B" is for a specimen initially the same as that of test "A", but which was loaded prior to the test to produce a prestrain of 5%. The uniaxial stress experiment "B" shown in Fig. 3.14 was then performed. We note that although the elastic limit of the material is far higher, the initial parabola still has a deformation mode index of $r = 4$, with a later sharper second-order transition to $r = 7$. Both of these tests were carried out at room temperature. It is interesting that the maximum stress was reached at the same total strain of 35%, despite the fact that test "B" was an interrupted experiment. Thus, we see that when a known prior deformation or preworkhardening is introduced, the parabolic large deformation generalization is still applicable.

Silver

Silver is of particular interest because of the stability of the initial parabola coefficient when the ambient temperature of the experiment is varied. Silver has a face-centered cubic crystal structure with a melting point T_m of 1,274° K, and with an experimental zero-point isotropic shear modulus $\mu(0)$ of 3,170 kg/mm². The 99.99% purity 0.564 in. specimen diameter uniaxial tension experiment of MᴄKᴇᴏᴡɴ and Hᴜᴅsᴏɴ (1937) shown in Fig. 3.15, provides a stable $r = 4$ parabola to 20% deformation followed by two double transitions to $r = 6$ and $r = 8$, as is characteristic of tensile testing. These data have been reduced to absolute zero for purposes of comparison by plotting $\left(\dfrac{\sigma}{1 - T/T_m}\right)^2$ vs ε.

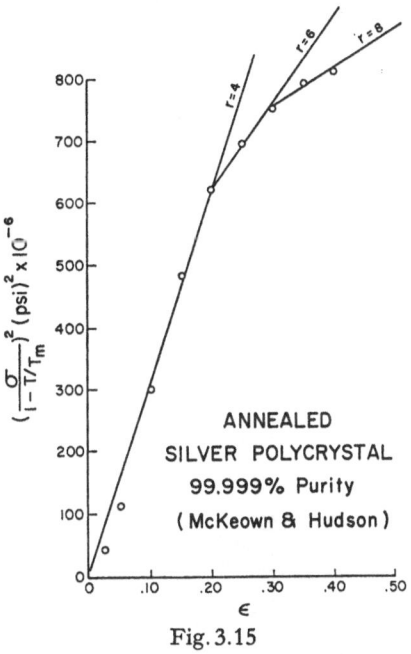

Fig. 3.15

CARREKER (1957) carried out an extensive series of polycrystalline tests in silver. His uniaxial stress experiments were tensile measurements in 0.020 in. diameter specimens of 99.97+% purity silver. The test temperatures varied from 20° K to 1,173° K; i.e., $T/T_m = 0.016$ to $T/T_m = 0.960$. Of particular interest in the present study is the effect of the differences in grain size produced by the use of three annealing temperatures: 973° K, 1073° K, and 1173° K. These three annealing temperatures produced grain sizes of 0.017, 0.040, and 0.250 mm diameter, respectively, after thirty minutes at each temperature. CARREKER's (1957) silver data, therefore, provide the opportunity of studying the combined effect of annealing temperature and grain size on the present writer's large deformation generalization.

As may be seen from an examination of twenty-three tensile experiments in silver (Fig. 3.16) this change of conditions has very little influence on the large deformation mode index. Unlike similar experiments described in Fig. 3.8 in aluminum, the initial large deformation mode index does not change with temperature despite the fact that the maximum fractional melting point temperature in aluminum was $T/T_m = 0.36$, whereas in the silver data it reaches the very high value of $T/T_m = 0.960$.

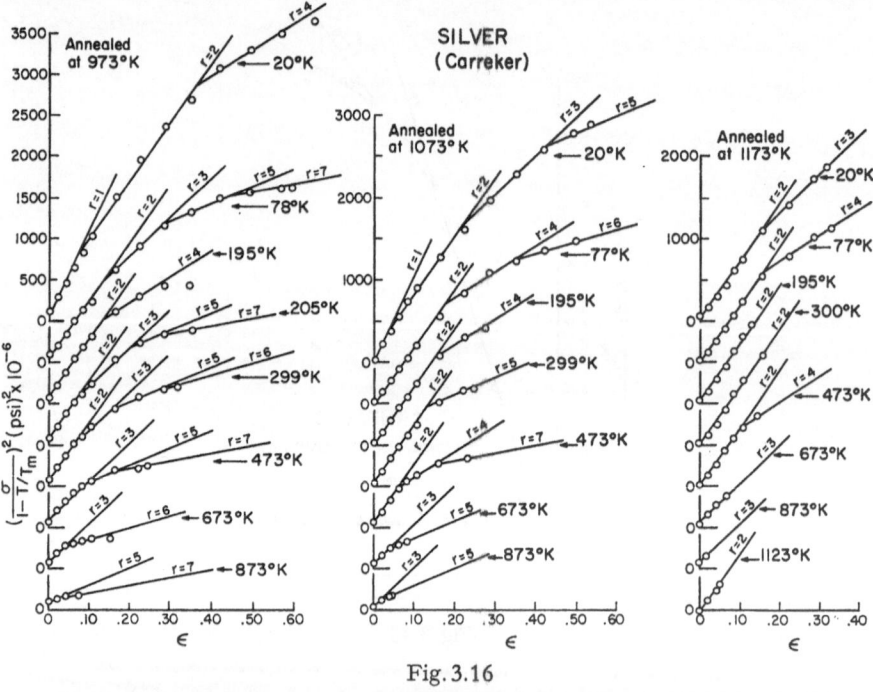

Fig. 3.16

In Chapter IV it will be shown that the stage III deformation of single crystals of lower purity aluminum do exhibit this stability of the initial parabola mode index. In some instances the mode index r remains unchanged to values of the fractional melting point temperatures as high as $T/T_m = 0.937$. (Dynamic polycrystalline aluminum gave an $r = 2$ mode index to $T/T_m = 0.980$.) The higher purity data of McKEOWN and HUDSON (1937) in Fig. 3.15 have an $r = 4$ initial parabola similar to the high purity aluminum data of Fig. 3.8 at 300° K. The differences in grain size of the silver data of Fig. 3.16 thus do have an effect upon the subsequent transition structure. For the highest annealing temperature, 1173° K, and largest grain size, 0.250 mm, the first transition for the three tests for which the deformation proceeded sufficiently far to produce a transition, occurred in the vicinity either of the fourth transition strain (11.5%) or of the fifth transition strain (16.3%). The strain at which the first transition occurs at the lower annealing temperatures is more variable, although for each individual test the strain at the second-order transition is consistent with those of the distribution predicted from aluminum.

In these data of CARREKER (1957) one may also note the predominance of double second-order transitions characteristic of tensile tests. These data, which are reduced to absolute zero through the use of a $\left(\dfrac{\sigma}{1 - T/T_m}\right)^2$ vs ε plot, were given by CARREKER in the form of reduced stress versus logarithmic strain; i.e., "true" stress versus "true" strain. Since CARREKER characterized the manner in which the reduction of the data was made, it was possible for the present writer to "detrue" the data for a comparison with the nominal parabolic uniaxial stress-strain generalization.

Copper

Copper is one of the polycrystalline elements whose finite amplitude wave propagation behavior has been studied extensively by the writer (BELL, 1963a, 1964; BELL and WERNER, 1962). As was shown in Figs. 2.24 and 2.25, diffraction grating wave propagation studies in this polycrystalline solid revealed that the one-dimensional finite amplitude wave propagation theory of TAYLOR (1942), VON KARMAN (1942), and RAKHMATULIN (1945) is applicable, having a parabolic stress-strain function which, at room temperature, has a deformation mode index of $r = 4$. The quasi-static uniaxial experiments in copper are sufficiently extensive so that the effects of ambient temperature, purity, type of test, initial work hardening, annealing temperature, and grain size may all be considered in a single solid. Uniaxial stress experiments exhibiting a variation of these parabolas are shown in Figs. 3.17, 3.18, and 3.19.

In Fig. 3.17 are shown two tensile tests of McKEOWN and HUDSON (1937) in 0.54 in. diameter, 99.99% purity copper at 300° K. These tensile tests are identical to those for gold described in Fig. 3.14; they were conducted in the same laboratory. Both specimens were initially annealed for thirty minutes at 923° K. Experiment "A" has an $r = 4$ parabola to the vicinity of the fourth transition strain at 11.5% where it undergoes a double transition to $r = 6$, followed by a second transition to $r = 9$ at a strain somewhat below the sixth transition strain of $27^0/_0$. The initial parabola large deformation index of $r = 4$ for this test at room temperature was the same as the dynamic value obtained for the majority of the finite amplitude wave propagation experiments in copper.

Also shown in Fig. 3.17 is an uniaxial tension experiment of Mc-KEOWN and HUDSON (1937) in a copper specimen prestressed $5^0/_0$ before the experiment was performed. The deformation is again parabolic, but

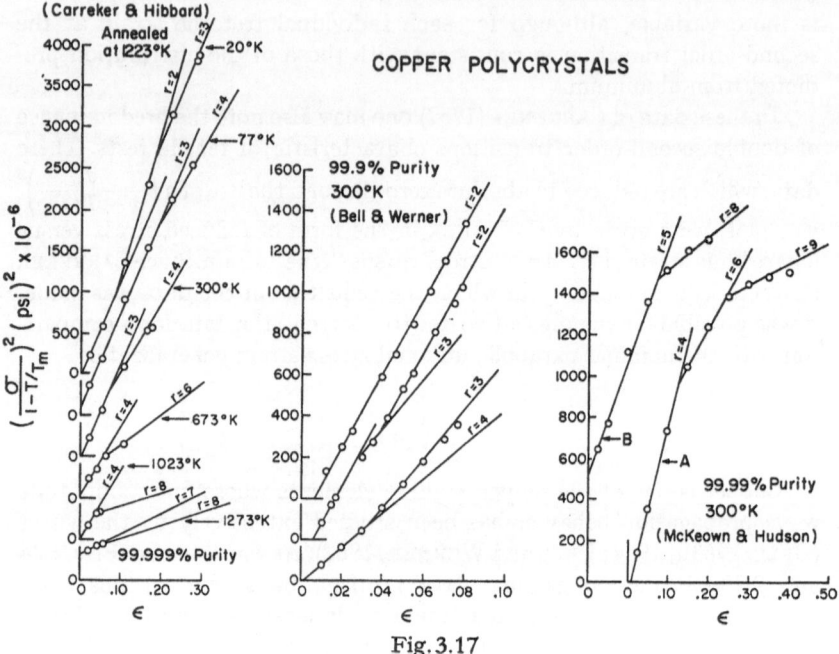

Fig. 3.17

the initial parabola now has a deformation mode index of $r = 5$ with a sharper transition to $r = 8$ at the third transition strain of $7.5^0/_0$. Several years ago BELL and WERNER (1962), in describing the results of finite amplitude wave propagation experiments in annealed copper, noted that a few experiments exhibited wave speeds which were smaller than those of the majority of the measurements. It may now be seen that these experiments had a parabolic dynamic stress-strain function with a deformation mode index of $r = 5$ rather than the usual value of $r = 4$. Therefore, it is interesting to note in the test of McKEOWN and HUDSON (1937) a similar multiple-valued mode index r for the initial parabola coefficient at room temperature.

Shown, too, in Fig. 3.17 are three quasi-static uniaxial compression experiments in 1 in. diameter, 3 in. long lubricated specimens. The experiments were performed by BELL and WERNER (1962) for comparison with the dynamic data. These wave propagation experiments which provided a predominantly $r = 4$ mode index were for a 99.9% purity, fine-grained polycrystal annealed at 867° K. One BELL and WERNER

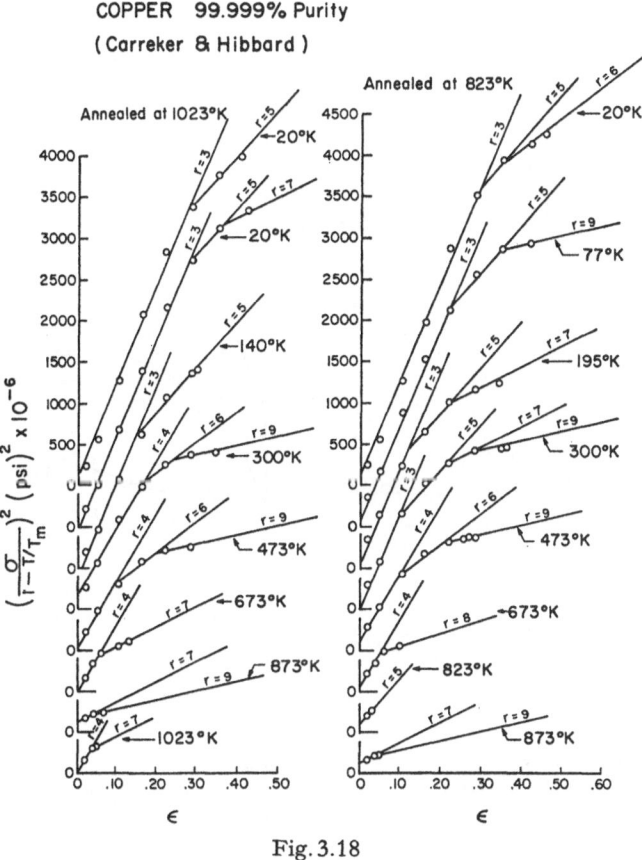

Fig. 3.18

quasi-static experiment of Fig. 3.17, which was also annealed at 867° K, has the same initial parabola of $r = 4$ with an upward transition to $r = 3$ at a strain slightly greater than 3%. The dynamic studies in copper of BELL and WERNER (1962) were all for maximum strains of less than 3%.

Two lubricated compression experiments of BELL and WERNER at the same annealing temperature of 867° K, also shown in Fig. 3.17, provided higher initial parabolas with large deformation mode indices of $r = 2$. The remainder of the data in Figs 3.17, 3.18, and 3.19 are tensile experiments of CARREKER and HIBBARD (1957) in 99.99% purity copper wire specimens of 0.030 in. diameter for fractional melting points from $T/T_m = 0.015$ to $T/T_m = 0.905$. Five different annealing temperatures of 523, 623, 823, 1,023, and 1,223° K were used to produce

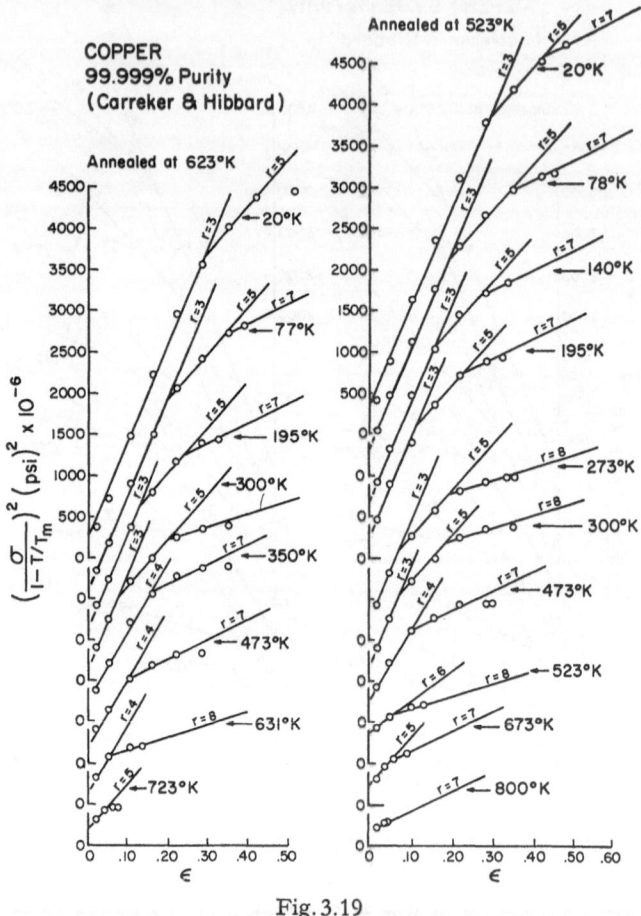

Fig. 3.19

average grain diameters of 0.012, 0.015, 0.030, 0.045, and 0.909 mm respectively. Irrespective of the ambient temperature of the test or the annealing temperature and corresponding average grain diameter, all of these 99.99% purity uniaxial tensile data of CARREKER and HIBBARD (1957) are parabolic. Predominant initial parabola large deformation mode indices are $r = 3$ and $r = 4$ until very high temperatures are reached, where initial values of $r = 5$ and $r = 7$ are observed.

All these data were presented in the CARREKER and HIBBARD paper in the form of reduced stress and logarithmic strain. Since the method of reduction was described in the paper, it was possible for this writer to recalculate the data to its original nominal form for a comparison with the generalized parabolic stress-strain function of Eq. (3.1). For comparative purposes, all of these data were reduced to absolute zero by considering $\left(\dfrac{\sigma}{1 - T/T_m}\right)^2$ vs ε plots. Apart from the obvious fact that these forty quasi-static uniaxial tension experiments provide an excellent agreement when the experimental data (circles) are compared with the predicted parabolas (solid lines), it is interesting to note that, as the ambient temperature of the test decreases to low values, the initial parabola proceeds to very large strains without undergoing a transition. At 20° K the strain of the first transition approximates the value of the sixth aluminum transition strain of 27%. One also notes that double second-order transitions, characteristic of tensile testing, predominate among these experiments, with sharper transitions occurring at very high temperatures. (Copper has a face-centered cubic crystal structure with an experimental zero-point isotropic shear modulus $\mu(0)$ of 5,080 kg/mm² and a melting point $T_m = 1{,}356°$ K.)

The different annealing temperatures of the CARREKER and HIBBARD (1957) data not only produced variations in the grain size but also variations in the initial elastic limit, as may be seen by the data in Fig. 3.17 annealed at 1,223° K in contrast to the data of Fig. 3.19 annealed at either 523° K or 623° K. Analysis of these forty-five uniaxial stress experiments in copper revealed that the writer's parabolic large deformation generalization with its mode and transition structure is applicable despite differences in grain size, purity, strain rate, specimen diameter, annealing temperature, prestress, initial elastic limit, and type of test; i.e., tension or compression.

Lead

Several years ago SPERRAZZA (1961; 1962a, b) performed a series of uniaxial stress finite wave propagation experiments on high purity lead in this writer's laboratory. The experiments were extremely difficult not only because of the problem of polishing lead so that 30,000 lines per inch diffraction gratings might be ruled on the surface, but also because at room temperature (300° K) which is half of the melting point temperature ($T_m = 600°$ K), lead polycrystals consist of a series of very large individual grains. The low zero-point isotropic shear modulus of $\mu(0) = 750$ kg/mm² for lead provided an equally low zero-point parabola coefficient with correspondingly small wave speeds.

Because of the large grain size, measurements had to be made on the individual grains in the polycrystalline matrix and large amounts of data had to be averaged to obtain wave speeds and maximum strains.

As was shown in Chapter II, the one-dimensional finite amplitude wave theory of TAYLOR (1942), VON KARMAN (1942), and RAKHMATULIN (1945) applied with a governing parabolic stress-strain function whose large deformation mode index is $r = 4$. The maximum average strain in SPERRAZZA's (1961; 1962a, b) tests was approximately 2.6%. It is interesting, therefore, to compare SPERRAZZA's quasi-static uniaxial compression experiment (Fig. 3.20) with this dynamic result. These quasi-static 3 in. long, 1 in. diameter specimens were lubricated.

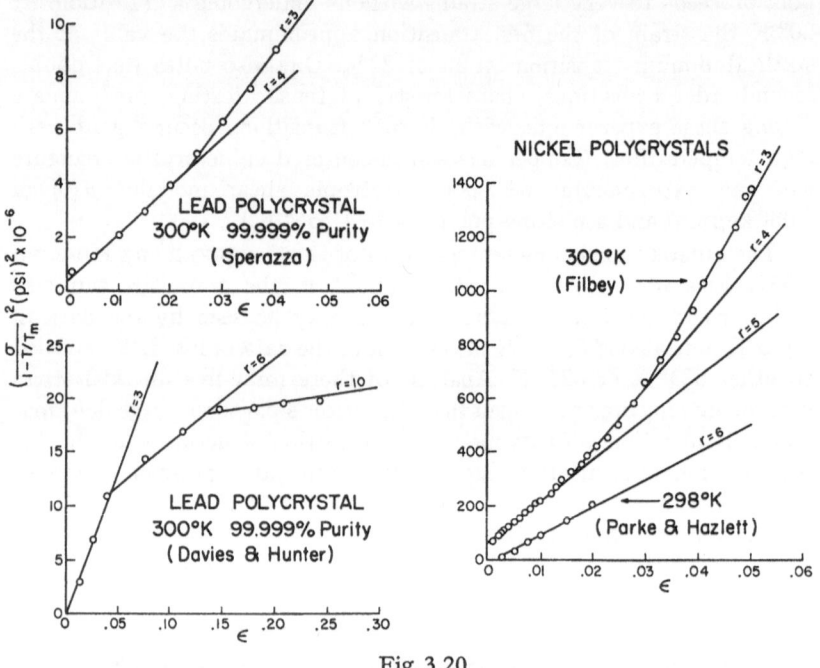

Fig. 3.20

As in the other metals thus far considered, the quasi-static deformation of lead not only is parabolic, but it is consistent with predicted parabolas from Eq. (3.1). The initial parabola of SPERRAZZA's uniaxial compression test has a mode index of $r = 4$, as in his dynamic data. This correspondence between dynamic and quasi-static stress-strain functions was, of course, noted by SPERRAZZA (1961; 1962a, b). In this quasi-static uniaxial compression experiment there is a unit transition from $r = 4$ to $r = 3$, as is characteristic of compression tests.

Also shown in Fig. 3.20 is an uniaxial tensile test of DAVIES and HUNTER (1963) in polycrystalline lead.

Nickel

Nickel at room temperature is a face-centered cubic crystal with a zero-point isotropic shear modulus $\mu(0)$ of 8,600 kg/mm² and a melting point, T_m, equal to 1,725° K. FILBEY (1965) in a series of finite wave propagation experiments in 1 in. diameter commercial purity polycrystalline nickel annealed at 955° K (described in Chapter II above), found that the finite wave theory of TAYLOR (1942), VON KARMAN (1942), and RAKHMATULIN (1945) applied. From such wave speed measurements, FILBEY determined a dynamic stress-strain function which was found to be parabolic. The computed parabola coefficient was found to be only 1% above the predicted value for an initial parabola large deformation mode index of $r = 5$. FILBEY's (1963) quasi-static uniaxial compression experiment, shown in Fig. 3.20, has an initial parabola coefficient with the same mode index $r = 5$ to the first transition strain at $\varepsilon_N = 1.5\%$, where a unit upward transition to $r = 4$ occurs, followed by a second upward transition to $r = 3$ at approximately 3%.

Also shown in Fig. 3.20 is a quasi-static uniaxial tension experiment of PARKER and HAZLETT (1953) in nickel.

Molybdenum

A polycrystalline molybdenum experiment of JAFFEE, MAYKUTH, and DOUGLASS (1961) was shown in Fig. 3.13. A far more extensive study of the uniaxial stress large deformation of this interesting body-centered cubic element was carried out by CARREKER and GUARD (1956) for the purpose of studying the deformation as a function of temperature and strain rate. Their experiments are uniaxial tension measurements in 99.95% purity molybdenum over a temperature range of 77° K to 1,813° K ($T/T_m = 0.027$ to $T/T_m = 0.63$). The specimens were 0.030 in. diameter wires with three different annealing temperatures and grain sizes. Specimens at 25° K, like the lower temperature Ta data of BECHTOLD, WESSEL, and FRANCE (1961) not only do not provide parabolic large deformation but they exhibited an essentially horizontal deformation after the elastic yield point was reached. However, 17 of CARREKER and GUARD's (1956) molybdenum experiments at ambient temperatures from 218° K to 1,809° K provided parabolic deformation consistent with the present writer's generalization, Eq. (3.1). Of particular interest in molybdenum are the extremely high elastic limits encountered, together with the abrupt increase in strain at the yield point, which

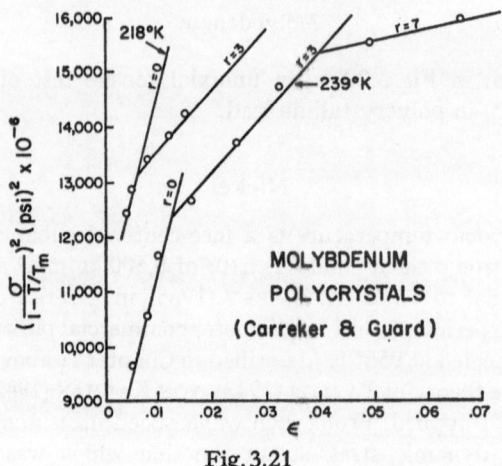

Fig. 3.21

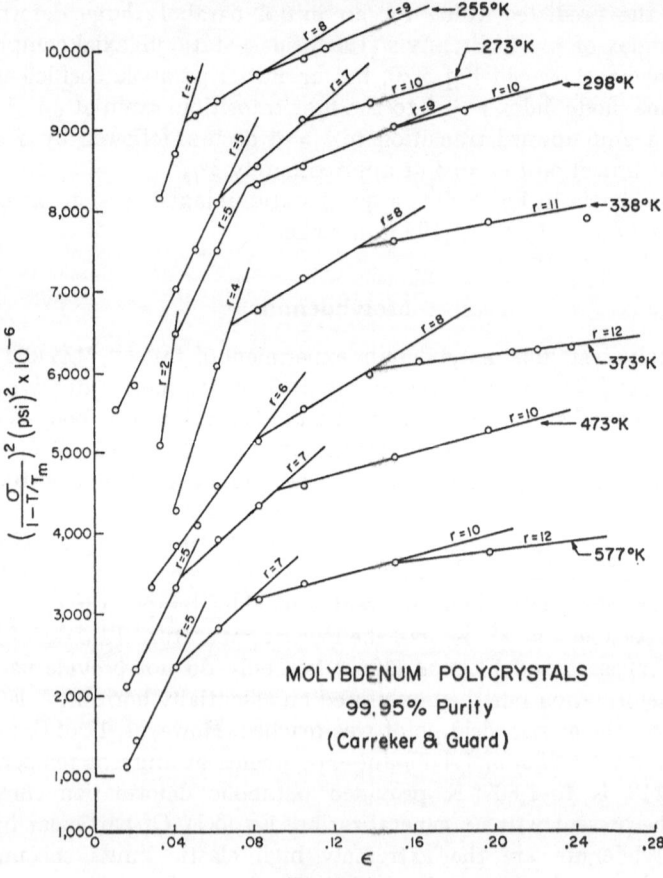

Fig. 3.22

characterizes both this body-centered cubic solid and iron, which latter
material will be discussed below.

These CARREKER and GUARD uniaxial tensile experiments are shown
in Figs. 3.21, 3.22, and 3.23. The tests above 673° K were carried out

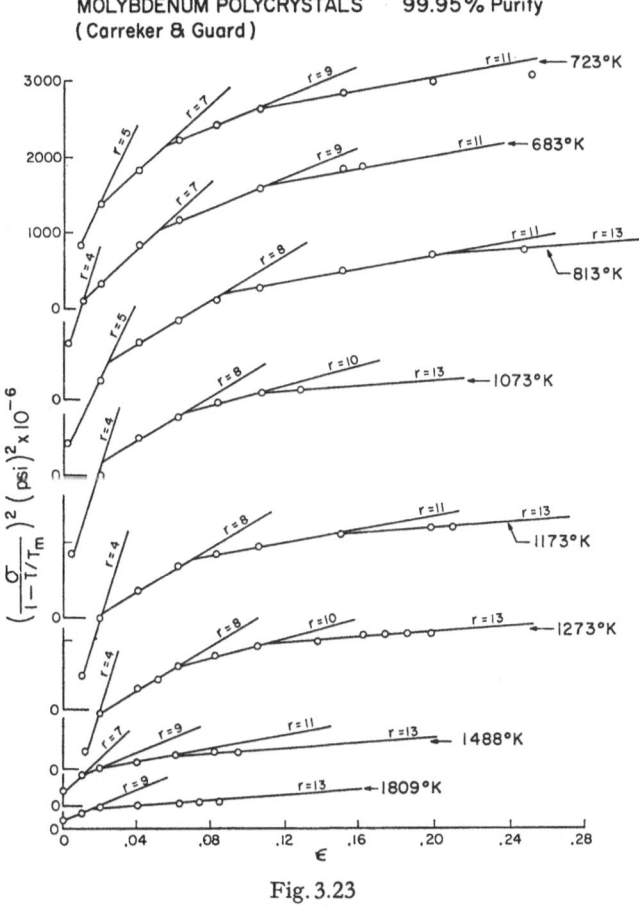

Fig. 3.23

in a vacuum to prevent oxygenation. The tested specimens were of three
gauge lengths, 2 in., 5 in., and 7 in. depending upon the experimental
conditions. This provided aspect ratios of 66 to 233. These data are
presented by CARREKER and GUARD in the form of reduced stress and
logarithmic strain, "true" stress — "true" strain, and had to be "detrued"
by the writer for comparison with the nominal stress-strain function of
the parabolic large deformation generalization.

It is of great interest that despite complexity introduced by the high elastic limit and subsequent discontinuous yielding, the experimental data (circles) are found to agree with predicted parabolas (solid lines) in the $\left(\dfrac{\sigma}{1 - T/T_m}\right)^2$ vs ε plots of Figs. 3.21, 3.22, and 3.23. All of these transitions either are double or are sharper, as is characteristic of tensile experiments in the other solids. The initial large deformation mode index r is zero below room temperature and remains at either $r = 4$ or $r = 5$ until ambient temperatures of $1,488°$ K and $1,809°$ K are reached, where initial parabola indices of $r = 7$ and $r = 9$ are observed.

Niobium (Columbium)

This crystalline solid is body-centered cubic. It has a zero-point isotropic elastic shear modulus $\mu(0) = 3,900$ kg/mm² and a melting point of $T_m = 2,743°$ K. Shown in Figs. 3.24 and 3.25 are six uniaxial tension experiments in columbium (niobium) of ENRIETTO, SINCLAIR, and WERT (1960). As may be seen from examining $\left(\dfrac{\sigma}{1 - T/T_m}\right)^2$ vs ε plots, the experimental data (circles) are in very good agreement with the prediction from the present writer's parabolic large deformation generalization (solid lines).

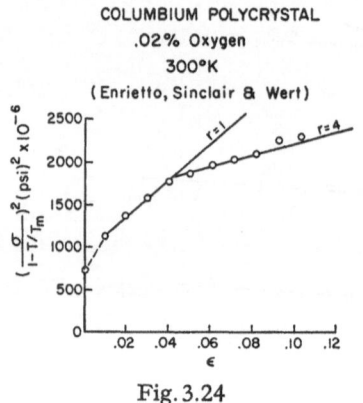

COLUMBIUM POLYCRYSTAL
.02% Oxygen
300°K
(Enrietto, Sinclair & Wert)

Fig. 3.24

The crystalline solid, niobium, is very sensitive to the presence of oxygen at higher temperatures. The experiments shown in Fig. 3.25 were annealed at $2,373°$ K for eleven hours to reduce the oxygen content to 0.001%. The subsequent experiments, shown in Fig. 3.25, exhibited an interesting behavior as the temperature was increased. At $300°$ K the parabola sequence is $r = 1$, to $r = 4$, to $r = 6$, to $r = 10$. As in other solids when the temperature is increased to $473°$ K, the initial parabola is lower, having an index of $r = 2$ followed by a double transition to $r = 4$ and $r = 6$, and then a transition to $r = 10$. When the temperature is increased to $673°$ K, one sees a reversal of the usual behavior and a much higher stress-strain function than at $473°$ K; at $673°$ K the successive deformation mode indices are $r = 1$ to $r = 3$ to $r = 8$. As the temperature increases to $773°$ K, the parabolas again become smaller with indices of $r = 1$ to $r = 4$ to $r = 9$; and at $873°$ K the indices further decrease to $r = 2$ to $r = 4$ to $r = 10$. This behavior may readily be seen in Fig. 3.25 since the experimental data are all calculated to absolute zero. These experimental data of ENRIETTO, SINCLAIR, and WERT (1960) also had to be "detrued" to the original nominal form for comparison with the present writer's generalized parabola.

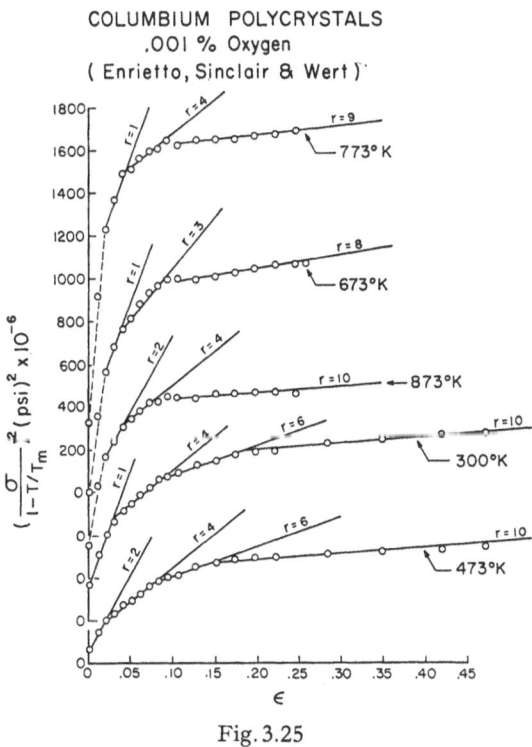

COLUMBIUM POLYCRYSTALS
.001 % Oxygen
(Enrietto, Sinclair & Wert)

Fig. 3.25

The experiment shown in Fig. 3.24 is for a specimen annealed for five minutes in a vacuum at 1,973° K, resulting in a solid with a 0.02% oxygen content. Only one of the five experiments is shown here; i.e., that at 300° K. As may be seen, the deformation is still consistent with the writer's parabolic generalization; and, indeed, the parabola sequence is the same as that for the experiment at 300° K of Fig. 3.25 with a 0.001% oxygen content ($r = 1$ to $r = 4$), with this transition occurring at the second transition strain of $\varepsilon_N = 4.2\%$.

The entire stress-strain curve for the niobium (columbium) with the higher content of oxygen lies far above that for the low oxygen content; i.e., the elastic limit of the former is $2\frac{1}{2}$ times higher than the latter.

Thus, once again one sees demonstrated the fact that the large deformation behavior is independent of the magnitude of the initial elastic limit of the material.

Chromium

Chromium is another body-centered cubic crystal structure for which some quasi-static uniaxial data is available. The experiments shown in Figs. 3.26 and 3.27 are uniaxial compression measurements of MARCIN-KOWSKI and LIPSITT (1962). They were performed in specimens with a 0.095 mm mean grain diameter at strain rates of $8.28 \times 10^{-5} \sec^{-1}$. The experiment shown in Fig. 3.26 at 473° K was an interrupted test carried out for the purpose of studying the strain-age-hardening phenomenon. A deformation mode index of $r = 4$ governs most of the deformation, but between 2% and 3%, unit transitions from $r = 4$ to $r = 3$ and $r = 3$ to $r = 4$ are observed. It is interesting that these transitions occur before the first test interruption, which was made at a strain in excess of 4%.

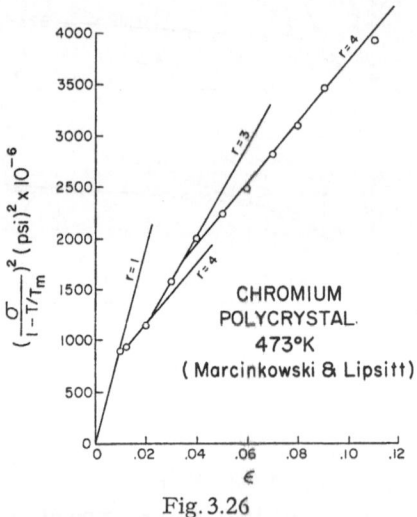

Fig. 3.26

The experiments shown in Fig. 3.27, like the data of Fig. 3.26, were plotted as nominal stress vs nominal strain. However, the data of Fig. 3.27 at the five different ambient temperatures shown, between 304° K and 673° K, were plotted in terms of the ratios of the instantaneous stress to the stress found at 3% plastic strain. It is not possible, therefore, to compare these data with specific parabolas because this ratio eliminates both the temperature term $1 - T/T_m$ and the zero-point isotropic shear modulus, $\mu(0)$ of Eq. (3.1). If the mode index r were to differ, variations of slope could occur. Despite the differences in the elastic limits of the various temperatures observed in the data of

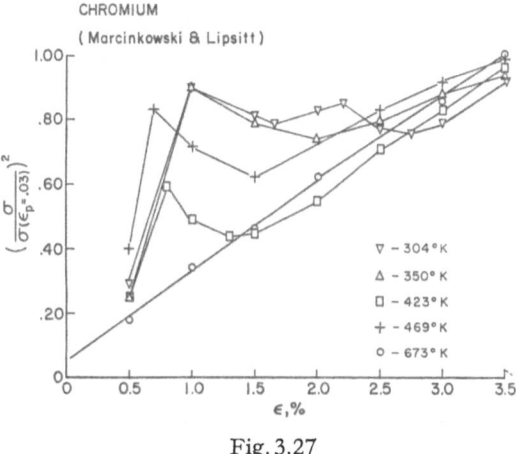

CHROMIUM

(Marcinkowski & Lipsitt)

Fig. 3.27

Fig. 3.27, one sees that as the deformation proceeds for the different temperatures, the $\left(\dfrac{\sigma}{\sigma\,(\varepsilon_p = .03)}\right)^2$ vs ε does provide the linearity characteristic of a parabolic deformation.

The observed parallelism of these curves at the different temperatures demonstrates that the parabolic deformation in chromium has the same temperature dependence as the other elements considered above. Chromium has a melting point T_m of 2,163° K and a calculated zero-point isotropic shear modulus $\mu(0)$ of 9,969 kg/mm² , both of which are far above the values for aluminum.

Iron

Iron has a body-centered cubic structure with a $\mu(0)$ of 8,700 kg/mm² and a melting point T_m of 1,806° K. From the point of view of technology, its behavior, and in particular the behavior of steel, is of major interest. Two interesting tensile experiments taken from the classical work of TAYLOR and QUINNEY (1931) are shown in Fig. 3.28. One experiment is for annealed mild steel and the second experiment is for decarburized mild steel. Both experiments are shown in $\left(\dfrac{\sigma}{1 - T/T_m}\right)^2$ vs ε plots.

The annealed mild steel experiment exhibits the high elastic limit and discontinuous yield phenomenon which characterizes this solid.

Nevertheless, the subsequent large deformation follows an $r = 2$ parabola, consistent with the present generalization. The decarburized mild steel specimen with a much lower elastic limit, as shown in Fig. 3.28, also has an experimental behavior in agreement with the writer's parabolic generalization. It is interesting that this *tensile* test in iron, unlike

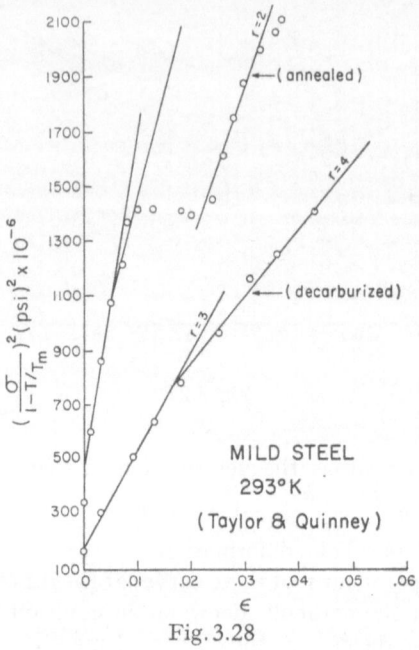

Fig. 3.28

similar data in other solids, undergoes a *unit* transition from $r = 3$ to $r = 4$ at the first transition strain of 1.5%. A series of compression experiments carried out by the writer in annealed ingot iron are described in Chapter VI. As will be seen these data also are in very close agreement with the present parabolic generalization, despite the magnitude of the initial discontinuous yield at the elastic limit.

Yttrium

Yttrium has a hexagonal crystalline structure with a zero-point isotropic elastic shear modulus $\mu(0)$ equal to 2,830 kg/mm², not very different from that of aluminum. Yttrium's melting point of $T_m = 1,763°$K is far above that for aluminum. Fig. 3.29 shows a $\left(\dfrac{\sigma}{1 - T/T_m}\right)^2$ vs ε plot for an uniaxial tension experiment in polycrystalline yttrium of SIMMONS (1961), in which it is found that the deformation is parabolic, with parabola coefficients in agreement with prediction from the present writer's generalization. SIMMONS' experiments in yttrium at room temperature consider the influence upon the uniaxial stress-strain function of the addition of a large number of additives. In Chapter VIII on binary combinations these data will be compared with the unalloyed experiment shown in Fig. 3.29.

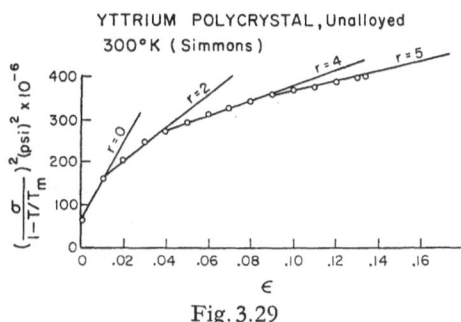

Fig. 3.29

Summary

In presenting the σ^2 vs ε polycrystalline uniaxial stress experimental data in the present chapter, little reference has been made to the expressed reasons for which each experimentalist performed his experiment. If such details were to be included in describing the large deformation polycrystalline experiments of the present chapter, or for the several hundred large deformation single crystal experiments described in the next chapter, this monograph would have to be expanded to several volumes. Despite the variety of atomistic hypotheses and continuum mechanics hypotheses which motivated their obtaining of these data, one sees that the data may be correlated into a single large deformation generalization.

Perhaps the variable of greatest interest to the majority of the experimentalists whose data have been presented, has been the strain rate $\dot{\varepsilon}$ from whose study they hoped to determine the viscous properties of metals. The strain rates of polycrystalline data presented here vary from $\dot{\varepsilon} = 10^{-8}$ sec^{-1} to $\dot{\varepsilon} = 1$ sec^{-1} for uniaxial stress quasi-static experiments and from $\dot{\varepsilon} = 1$ sec^{-1} to $\dot{\varepsilon} = 10^4$ sec^{-1} for uniaxial stress finite amplitude wave propagation experiments. In this range of strain rate of 12 orders of magnitude, the same generalized linearly temperature-dependent parabolic nominal stress-strain function, Eq. (3.1), is applicable.

For the 19 polycrystalline elements and for the 5 binary combinations, numerous mixtures, and low percentage alloys discussed in Chapter IV and Chapter VIII, one sees that alterations in the strain-time history of the deformation do not in any instance produce a departure from the linearly temperature dependent parabolic large deformation generalization. Within the framework of the generalization, differences in the transition structure at critical strains are observed when very low strain rate experiments are compared with very high strain rate experiments. The main problem which results from a variation of loading histories is thus seen to be one of determining stability properties rather than viscous behavior.

Of equal importance with variations in the strain-time or stress-time history of the deformation is the purity, ambient temperature, and type of test (tension or compression), all of which influence the discrete large deformation mode index r of the initial parabola and the subsequent transitions. In the large deformation generalized stress-strain function of Eq. (3.1), the melting point T_m, the test temperature T, the zero-point isotropic shear modulus $\mu(0)$, and the universal dimensionless constant B_0, are all specified. The only undetermined parameter is the mode index r. In a σ^2 vs ε plot, as was shown in Fig. 3.1, a unit change in the integer r produces an easily recognized, discrete change in slope.

In obtaining the polycrystalline data described in the present chapter, as is also true of the literature data of single crystals of Chapter IV and elastic shear moduli data of Chapters V and VI, *none* of the experiments in the several hundred papers examined has been excluded. Thus, the experimental large deformation generalization has been submitted to the severest of tests: the test of describing the major portion of the extant literature. The simplicity and accuracy of the uniaxial stress measurement, as was pointed out above, is such that any meaningful generalization must be able to include the major portion of the extant experimental data. That the present writer's parabolic generalization accomplishes this for polycrystalline data has been amply demonstrated in the present chapter.

The large deformation generalization of Eq. (3.1), discovered by the writer from diffraction grating studies in finite amplitude wave experiments, also is seen to be applicable to quasi-static experiments despite the variations in crystal structure, strain rate, purity, grain size, annealing temperature, prior work hardening, elastic yield point, test temperature, zero-point isotropic shear modulus, specimen geometry (diameters from 0.020 in. to 1.00 in., etc.), type of test, and melting point.

Chapter IV

Single Crystal Uniaxial Stress Experiments

An uniaxial stress experiment in a single crystal, whether dynamic or quasi-static, is complicated by the fact that such solids are anisotropic. Large deformation occurs as shear on preferred planes in specified directions. In a quasi-static experiment, the two axial measurements for the single crystal are the same as for the polycrystal, and in the field of large deformation may be determined with the same high degree of accuracy. Difficulties arise when the data from one quasi-static experiment are compared with the data from another. Unless the orientation of crystallographic axes with respect to the specimen axis is known, one may not meaningfully cross-compare data in any particular solid, let alone make comparisons between solids.

This means that to interpret the experimental data for large deformation uniaxial stress experiments in single crystals, x-ray diffraction measurements must be made to determine not only the initial orientation of the crystallographic planes with respect to the specimen axes but also their subsequent rotations during deformation. Given orientation data, kinematical considerations must supply a method of calculating resolved shear stress, resolved shear strain on the appropriate planes and in the appropriate directions. These additional x-ray measurements and their interpretation introduce new sources of error rendering single crystal data less reliable than polycrystal data. Single crystal data from uniaxial stress experiments have thus been averaged for comparison with prediction from the writer's large deformation generalization.

Unfortunately, few experimentalists who perform quasi-static uniaxial stress experiments in crystalline solids provide the original nominal stress and strain data upon which their calculations of resolved data are based; hence, it is not possible to recalculate their data using different resolving procedures. In recent years it has become the practice for many experimentalists in crystal physics to present computed resolved shear stress, resolved shear strain data only in tabulated form. Such tabulated data usually are confined to the listing of certain stated peculiarities of the test results which have significance with respect to some particular atomistic hypothesis of interest to that investigator.

A schematic resolved shear stress, resolved shear strain function for uniaxial stress experiments in high purity cubic single crystals is shown in Fig. 4.1.

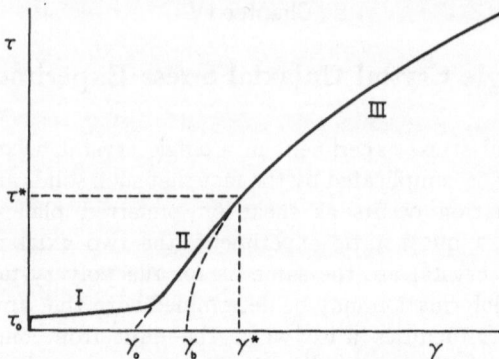

Fig. 4.1. A schematic diagram of the resolved deformation of a high purity single crystal, showing the three stages of deformation

Of the three stages of deformation, that which has been most extensively studied is stage I, with particular emphasis on the region in the immediate vicinity of the elastic limit or critically resolved shear stress τ_0. Stage I, known as the region of "easy glide," occurs primarily in high purity cubic crystalline solids whose initial crystallographic orientation is not in one of the three corners of the stereographic triangle. The representation of the angles of the cylindrical specimen axis with respect to crystallographic planes and directions customarily is given by crystal physicists in the form of a stereographic projection. The crystal is assumed to be located at the center of a reference sphere with its planes extended until they intersect the sphere. The details of this projection are given in standard texts such as BARRETT (1952). Extensive theoretical studies in the field of crystal imperfections have been made to interpret the observed properties of stage I behavior. In recent years numerous efforts have been made to extend such crystal imperfection theories to include the linear work-hardening stage II deformation. The large number of such atomistic theories which have been developed to describe and numerically account for this linear stage II behavior attests to the limitations of such atomistic hypotheses in unraveling the importance of each of a multiplicity of mechanisms in what is obviously a complicated physical situation.

The resolved stage III deformation of the single crystal with which this monograph is concerned is so admittedly complex that, until very recently, no efforts had been made to describe the large deformation behavior in terms of atomistic imperfection theories. When the purity of the cubic crystal is sufficiently low or when the initial orientation is in the vicinity of one of the corners of the stereographic triangle (face-centered cubic single crystals in corner orientations have a specimen axis along a side of the cube or along a face or volume diagonal), no stage I or stage II deformation is observed. The present writer has shown (BELL, 1964) that the stage III deformation was unaffected by variations in the length of the stage I deformation or the slope, θ_{II}, of the stage II deformation. Indeed, the stage III behavior was unaffected when both stage I and stage II were completely absent.

In that paper (BELL, 1964) and in a succeeding paper (BELL, 1965a) the writer showed that the resolved stage III deformation of over 300 experiments in face-centered cubic crystals of Al, Ag, Au, Ni, Pb, and Cu not only were parabolic but also had parabola coefficients quantitatively determinable *empirically* from the nominal polycrystalline generalization of Eq. (1.1) through the stress and strain ratios of the TAYLOR (1938) and BISHOP and HILL (1951) aggregate theory.

It was further shown (BELL, 1964, 1965a) that stage II deformation, when present, is analytically related to stage III deformation in a manner such that stage III parabola coefficients could be determined solely from the knowledge of stage II parameters when the stress τ^* at the intersection of the two stages of deformation is known. For the linear resolved stage II deformation one may write:

$$\tau = \theta_{II}(\gamma - \gamma_0) \tag{4.1}$$

where γ_0 is the intercept of the stage II deformation on the strain abscissa. For parabolic resolved stage III deformation one has

$$\tau = \beta(\gamma - \gamma_b)^{1/2} \tag{4.2}$$

where γ_b is the intercept of the stage III parabola on the strain abscissa and β is the parabola coefficient for the solid and temperature of interest. Equating stresses and first derivatives at τ^* after noting experimentally that stage II proceeds smoothly into stage III, one obtains Eqs. (4.3) and (4.4) respectively.

$$\theta_{II}(\gamma^* - \gamma_0) = \beta(\gamma^* - \gamma_b)^{1/2} \tag{4.3}$$

$$\theta_{II} = {}^1\!/_2\beta(\gamma^* - \gamma_b)^{-1/2} \tag{4.4}$$

where γ^* is the strain at τ^*. The parabola coefficient β of the stage III deformation and parabola intercept γ_b, on the strain abscissa thus may be determined in terms of stage III parameters:

$$\beta = \sqrt{2\tau^* \theta_{II}} \tag{4.5}$$

$$\gamma_b = \frac{\gamma^* + \gamma_0}{2}. \tag{4.6}$$

Calculations of parabolic stage III (circles) by means of Eqs. (4.5) and (4.6) are shown in Fig. 4.2 for the room temperature high purity copper data of several experimentalists.

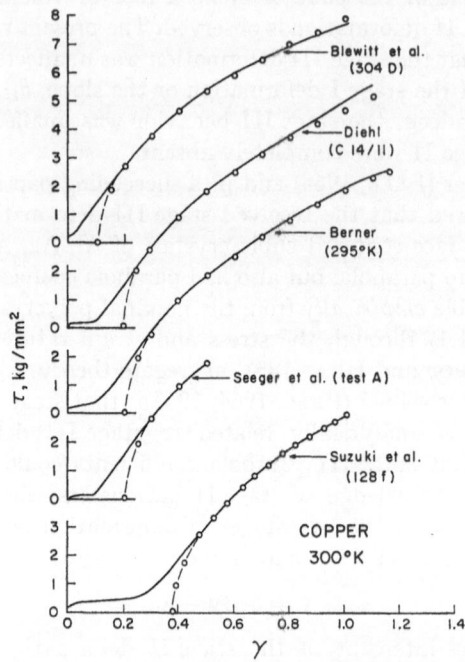

Fig. 4.2. High purity copper tension tests (solid lines) compared with calculated stage III (circles) from Eqs. (4.5) and (4.6)

Parabola coefficients determined by means of Eq. (4.5) are designated in the remainder of this chapter as β(II) to indicate that they were calculated from stage II data.

Parabola coefficients also may be determined directly from the resolved stage III data itself by squaring both sides of Eq. (4.2), which furnishes Eq. (4.7)

$$\beta = \left[\frac{\tau^2}{(\gamma - \gamma_b)}\right]^{1/2}. \tag{4.7}$$

When parabola coefficients are determined by such direct measurements of the stage III data they are designated as β(III).

A comparison of averaged β_{ro}(II) and averaged β_{ro}(III) determinations from nearly 400 single crystal experiments in 6 face-centered cubic elements has established Eqs. (4.5) and (4.6) to within a fraction of a percent (BELL, 1961b, 1964, 1965a) (BELL and WERNER, 1962) (where β_{ro} is the zero-point parabola coefficient). KUHLMANN-WILSDORF

(1967) recently has proposed that the present writer's interrelation of stage II and stage III deformation in terms of Eqs. (4.5) and (4.6) may be described in terms of dislocation theory.

An observation of considerable importance, shown by the writer in earlier papers (BELL, 1961b, 1964, 1965a, b) (BELL and WERNER, 1962), is the fact that the parabola coefficient for the stage III resolved shear deformation of the cubic single crystal may be empirically determined from the introduction of the *nominal* parabolic stress-strain function, Eq. (1.1), from the uniaxial stress polycrystalline experiments, into the stress and strain ratios of the TAYLOR (1938) and BISHOP and HILL (1951) aggregate theory, Eq. (4.8).

$$\frac{\sigma}{\tau} = \overline{m} = \frac{\gamma}{\varepsilon}. \tag{4.8}$$

Using the principle of virtual work, TAYLOR, in 1938, showed that from a consideration of the behavior of the five independent slip systems of the total of twelve possible systems for face-centered cubic crystals, it was possible to predict the behavior of a face-centered cubic polycrystalline aggregate in terms of the stress and strain ratios of Eq. (4.8) when the resolved shear stress-strain function $\tau = f(\gamma)$ is specified. BISHOP and HILL in 1951 obtained the same results for face-centered cubic solids on the basis of a consideration of the maximum plastic potential. In both of these theoretical developments the same value of the dimensionless upper bound $\overline{m}$ was found; i.e., $\overline{m} = 3.06$. The predicted polycrystalline relation thus may be written as Eq. (4.9).

$$\sigma = \overline{m} f(\overline{m}\varepsilon). \tag{4.9}$$

Conversely, one may determine the single crystal resolved shear stress-strain function, given the polycrystalline stress-strain function. Thus the β's of Eqs. (4.5) or (4.7) for either $\beta(II)$ or $\beta(III)$, have the form of Eq. (4.10) upon introducing Eq. (1.1) into Eq. (4.8).

$$\beta = (2/3)^{r/2} \mu(0) \frac{B_0}{\overline{m}^{3/2}} (1 - T/T_m). \tag{4.10}$$

Thus, Eq. (4.2) for the stage III deformation of the face-centered cubic single crystal becomes Eq. (4.11).

$$\tau = (2/3)^{r/2} \mu(0) \frac{B_0}{\overline{m}^{3/2}} (1 - T/T_m) (\gamma - \gamma_b)^{1/2}. \tag{4.11}$$

In comparing one single crystal experiment with another reference is made to the zero-point parabola coefficient in Eq. (4.11); i.e., $\beta_{r0} = (2/3)^{r/2} \mu(0) \frac{B_0}{\overline{m}^{3/2}}$.

In the remainder of the present chapter the resolved shear single crystal deformation of 8 cubic crystals and, in Chapter VIII, the single crystal stage III deformation of 4 binary combinations will be examined in terms of the writer's large deformation generalization as expressed in Eq. (4.11). As will be shown below, the fact that cubic single crystals provide averaged resolved stage III large deformation in agreement with prediction, of course extends the generality of the distortional deformation behavior being considered in the present study.

It is curious, however, that when this comparison of polycrystals and single crystals is made, the ratios of Eq. (4.8) with $\overline{m} = 3.06$ are applicable not only to the face-centered cubic situation for which the aggregate theory was developed, but also to the body-centered cubic elements, iron and tantalum. That no change in $\overline{m}$ is observed empirically suggests that one must regard the stress and strain ratios of the aggregate theory as an upper bound with Eqs. (4.8) and (4.9) possessing more generality than is implied in its theoretical development for the face-centered cubic solid.

In accordance with kinematical considerations introduced by TAYLOR and ELAM (1923) and by GOLER and SACHS (1927) the resolved shear stress and resolved shear strain are given by Eqs. (4.12) and (4.13) for the primary slip system having the quantitatively largest resolved shear stress.

$$\cos \varphi = \frac{\cos \varphi_0}{1 + \varepsilon}, \quad \sin \lambda = \frac{\sin \lambda_0}{1 + \varepsilon} \tag{4.12}$$

$$\gamma = \frac{\cos \lambda}{\cos \varphi} - \frac{\cos \lambda_0}{\cos \varphi_0} \tag{4.13}$$

$$\tau = \sigma \cos \varphi_0 \cos \lambda.$$

Where φ is the angle between the specimen axis and the normal to the plane upon which shear occurs, and λ is the angle between the specimen axis and the shear direction; φ_0 and λ_0 designate the initial pre-deformation value of these angles.

When the specimen axis rotated to a position of symmetry such that the resolved shear stress became equal in two slip directions, TAYLOR and ELAM (1923) and GOLER and SACHS (1927) then assumed that the specimen axis would move along this symmetry line toward the point where the great circle through the primary and conjugate slip directions cuts it, and, upon reaching this intersection point, no further movement of the axis would take place. In this situation, known as double slip, Eqs. (4.12) and (4.13) were no longer used. The double slip equations which were introduced presumed equal contributions to the resolved shear deformation from each of the two slip directions. In the three corners of the stereographic triangle the number of initial slip systems is 4, 6, or 8, depending upon the particular initial orientation. In this situation multiple slip is presumed to occur throughout the deformation and no rotation of the specimen axis with respect to the crystallographic planes and directions is presumed to take place.

During single slip, the angles φ and λ of Eqs. (4.12) and (4.13) may be measured by means of x-ray diffraction as the deformation proceeds. However, the actual resolved shear stress and resolved shear strain are not directly measured, so one may not be certain that the proposed resolving procedures are correct even if the measured angles are consistent. The stage III deformation of interest in this monograph is well known to be a region of complex multiple slip, hence one wonders why in the literature of the past 40 years such data have been resolved in single slip.

The present writer's experimental use of the stress and strain ratios of the TAYLOR (1938) and BISHOP and HILL (1951) aggregate theory constitutes a method of examining resolved shear stress, resolved shear strain single crystal data independently of the resolving procedures described above (BELL, 1961b, 1963a, 1964, etc.). From the examination of several hundred single crystal stage III experiments in a number of crystalline solids this writer found that agreement with the parabolic generalization of Eq. (3.1) was obtained when the data were calculated for macroscopic single slip by means of Eqs. (4.12) and (4.13). When stage III single crystal data were resolved by means of the double slip equations of TAYLOR and ELAM (1923) or of GOLER and SACHS (1927) or when calculations were made for multiple slip in initial corner orientations, the resolved shear stress-strain curves in general did not agree with prediction. When these same experiments were recalculated for macroscopic *single* slip in terms of Eqs. (4.12) and (4.13), agreement with the parabolic generalization was obtained, thereby indicating that multiple slip resolving procedures for stage III deformation (BELL, 1961b, 1964, 1965a) (BELL and WERNER, 1962) do not apply to the physics of large deformation.

To explore this unexpected conclusion of the present writer, BELL and GREEN (1967) surveyed over 40 years of the single crystal literature to determine how many experimentalists had actually measured φ and λ by x-ray diffraction *during* the deformation when the specimen axis rotated across a symmetry line or lay initially in a multiple slip corner orientation. In the entire literature, 168 experiments were found (including 16 aluminum measurements made by BELL and GREEN) for which only 9, all before 1930, behaved in the manner assumed in the double slip calculations. A reexamination of the 9 has shown that 3 of these, the data of KARNOP and SACHS (1927), were not actual measurements but were a schematic representation of what was to be expected. Thus only 6 of 165 experiments for which measurements actually were made during deformation on both sides of the symmetry line, rotated in a manner prescribed by the double slip calculation. Five measurements were found which had initial corner orientations very close to the

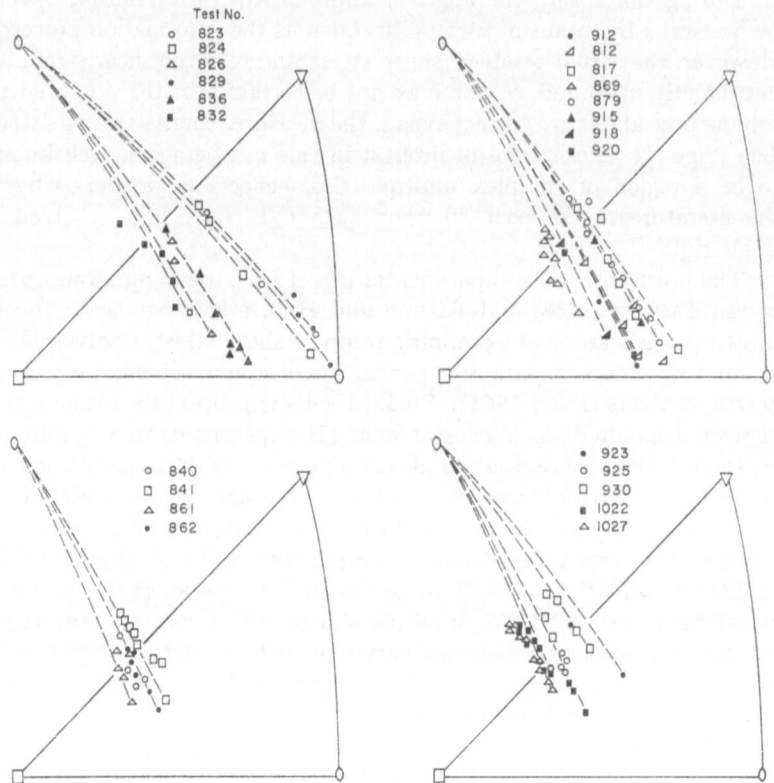

Fig. 4.3. X-ray diffraction measurements during deformation for single crystals whose axes rotated across the symmetry line

⟨100⟩ corner of the stereographic triangle where 8 slip systems are equally probable and where, in terms of multiple slip, no rotation is to be expected. All 5 of these experiments had specimen axis rotations away from this corner in precisely the manner to be expected in terms of the macroscopic single slip behavior of Eqs. (4.12) and (4.13).

In Fig. 4.3 are shown the specimen axis rotations during deformation for 14 of BELL and GREEN's (1967) aluminum experiments which crossed the symmetry line. For all of these data the axis continued to rotate linearly in single slip.

It was further shown by BELL and GREEN (1967) that if these x-ray diffraction measurements of φ and λ during deformation were introduced into Eqs. (4.12) and (4.13) then the values of τ and γ coincided with the single slip calculation based solely upon the initial measurements of λ_0 and φ_0, and differed from the double slip calculations assuming specimen axis rotation along the symmetry line. Such calculations (circles) made from the x-ray diffraction measurements are compared in Fig. 4.4 with calculated single slip and calculated double slip resolved shear stress, resolved shear strain (solid lines).

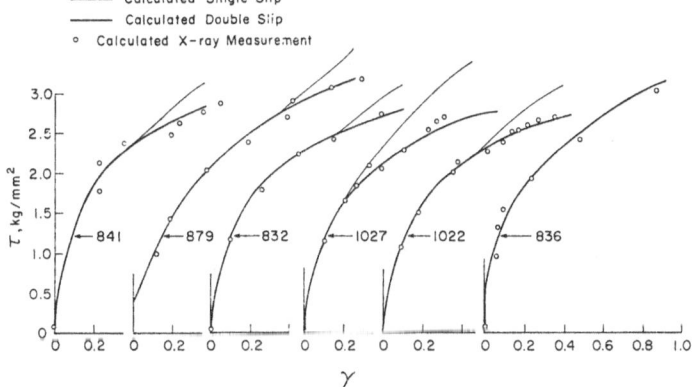

Fig. 4.4. Aluminum single crystal experiments resolved for single slip and for double slip, compared with prediction, Eqs. (4.12) and (4.13), from x-ray diffraction measurements during the deformation

Thus, despite the fact that from surface observations stage III deformation is known to be a region of complex multiple slip, the resolved deformation is given by the macroscopic single slip Eqs. (4.12) and (4.13), no matter where the specimen axis is initially located with respect to crystallographic planes and directions.

Observations have shown that $\mu/\theta_{II} \approx 300$ and is sensibly temperature independent. Over 200 experimental values of θ_{II} in 8 cubic metals are tabulated in Appendix I. An inspection of these data reveals that the scatter in measured values at any given temperature is at least $\pm 10\%$, and may be as high as $\pm 30\%$. However, comparison of averaged values at different fractional melting point temperatures T/T_m reveals that θ_{II} does not vary with temperature. Averaging all of the values of θ_{II} for each of the 8 metals, one obtains: Ag, 8.9 kg/mm²; Al, 8.4 kg/mm²; Au, 9.9 kg/mm²; Cu, 13.7 kg/mm²; Ni, 23.5 kg/mm²; Pb, 2.9 kg/mm²; Fe, 14.3 kg/mm²; and Ta, 9.1 kg/mm². The weighted average of the 193 fcc experiments in Appendix I furnishes $\dfrac{\mu(0)}{\theta_{II\,avg.}} = 348$.

[One notes that the two body-centered cubic elements, Fe and Ta, have a value $\left(\frac{\mu(0)}{\theta_{II}} = 611 \text{ and } 770\right)$ approximately twice that of the 6 fcc elements.] On the basis of the aggregate theory ratios of Eq. (4.8), if the function $f(\gamma)$ in Eq. (4.9) is linear, θ_{II} is given by Eq. (4.14).

$$\theta_{II} = \frac{\mu(0) B_0}{\overline{m}^2}. \qquad (4.14)$$

With the value of $\overline{m} = 3.06$ and $B_0 = 0.0280$ we, therefore, see that $\mu(0)/\theta_{II} = \overline{m}^2/B_0 = 334$ which may be compared with the experimental value of 348. Introducing θ_{II} from Eq. (4.14) into Eq. (4.1) furnishes Eq. (4.15) for fcc stage II deformation.

$$\tau = \mu(0) \frac{B_0}{\overline{m}^2} (\gamma - \gamma_0). \qquad (4.15)$$

Combining Eqs. (4.10) and (4.15) in Eq. (4.5) furnishes Eq. (4.16) for the stress τ^* at the intersection of stage II and stage III deformation.

$$\tau^* = \frac{(2/3)^r \mu(0) B_0}{2\overline{m}} (1 - T/T_m)^2. \qquad (4.16)$$

Aluminum

We begin this comparison of resolved single crystal experimental data with the parabolic stage III generalization of Eq. (4.11) by describing a series of dead-weight tension and compression quasi-static uniaxial stress experiments in aluminum single crystals of three different purities. All these experiments, each of whose initial orientation of specimen axis is shown in Fig. 4.5, were carried out in the writer's laboratory.

The first group of experiments in this series is the dead-weight, constant stress rate uniaxial tension experiments on 99.16% aluminum single crystals. The data from these experiments, carried out by SHARPE (1966a, b, c) as part of his experimental study of the effect of grain boundaries on the Portevin - le Chatelier effect, are shown in Fig. 4.6.

The constant nominal stress rate in these experiments was identical with that of the polycrystalline aluminum experiments of Fig. 3.3; i.e., 5 psi sec^{-1}. These polycrystals also were of 99.16% purity with precisely the same specimen diameter and length. The single crystals were grown by the strain anneal technique.

The experimental apparatus in the single crystal measurements of Fig. 4.6 was the same as that of the polycrystalline experiments shown in Fig. 3.2. Every one of the 16 single crystal experiments is parabolic, as the comparison of the experimental data (circles) and prediction (solid lines) from Eq. (4.11), in the τ^2 vs γ plot of Fig. 4.6 demonstrates.

The initial parabola for all 16 tests is $r = 2$, as in the dynamic poly-crystal study. A second-order double transition to $r = 4$, characteristic of tension tests, occurs in nearly every instance. The aggregate theory ratios of Eq. (4.8) from which Eq. (4.11) was obtained from the poly-crystalline stress-strain function Eq. (3.1), provide a relation between

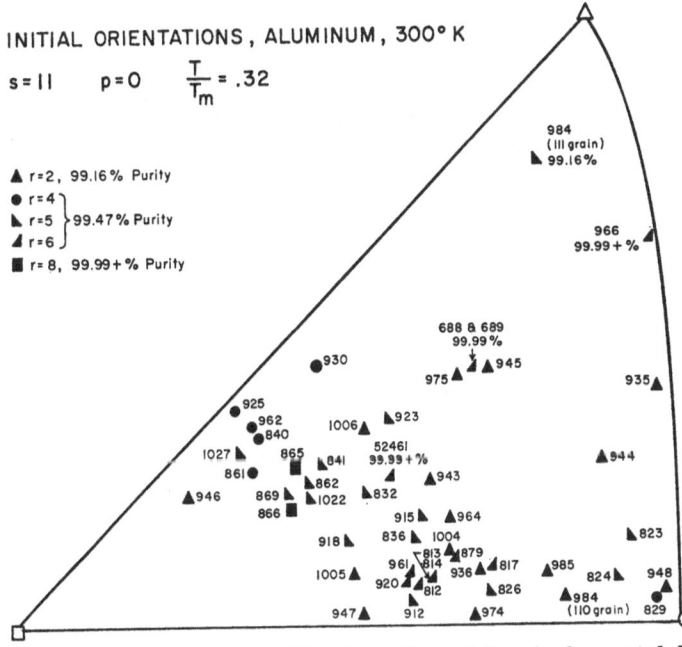

Fig. 4.5. The initial crystallographic orientations of the single crystal data of the writer shown in Figs. 4.6 to 4.10

single crystal shear strain γ and the polycrystalline nominal strain ε of $\gamma = \overline{m}\varepsilon$. Thus, the polycrystalline transition strains of $\varepsilon_N = 1.5$, 4.2, 7.5, 11.5, 16.5, and 27% become $\gamma_N = 4.6$, 12.9, 23, 35, 50.5, and 81.4%. The 99.16% aluminum single crystals of Fig. 4.6, governed by an initial parabola of $r = 2$, undergo the first transition at an average strain of 13%, which is almost exactly that predicted from the aggregate theory ratios for the same second transition strain, 12.9%.

The fact that single crystal data furnish initial parabola coefficients identical with those obtained from the dynamic compression wave propagation studies in polycrystals of this same purity has been noted before by this writer. The single crystal tension experiments of Fig. 4.6 had a variable strain rate of 10^{-5}, whereas the dynamic polycrystalline experiments described in Chapter II were in compression with variable strain rates 6 to 9 orders higher.

In understanding the large deformation generalization it is obviously of significance to note that the high strain rate single crystal or polycrystalline finite distortional behavior in wave propagation studies in general have large deformation mode indices r, correlating with the quasi-static single crystal behavior when purities are the same.

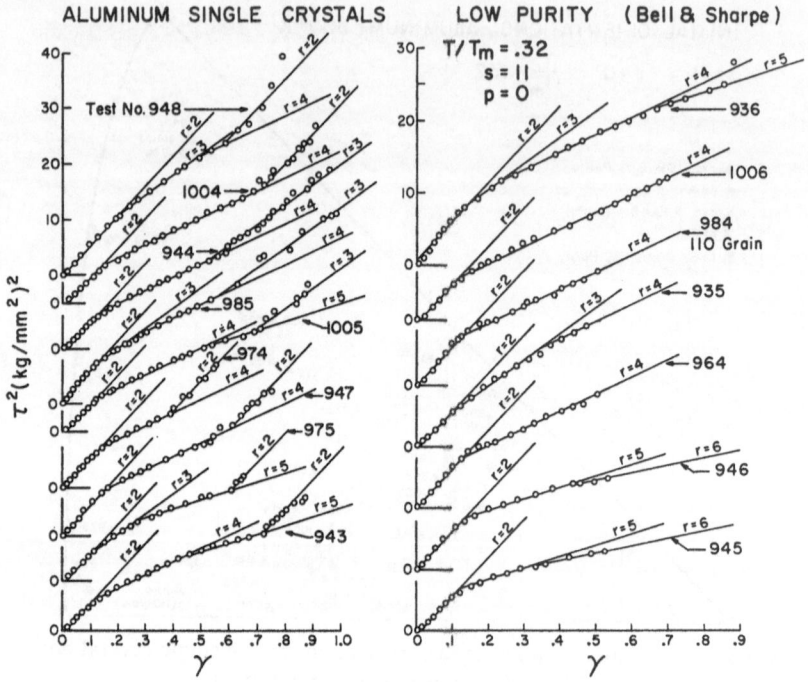

Fig. 4.6. τ^2 vs γ plots of low purity aluminum single crystals resolved in macroscopic single slip

A large number of both compression and tension quasi-static uniaxial stress experiments in 99.16% polycrystalline aluminum had initial parabolas of $r = 2$, $r = 3$, or occasionally $r = 4$, thereby demonstrating the fact that stability is a major question at low strain rates. Polycrystalline data with initial mode indices of $r = 2$ were shown in Fig. 3.4.

The resolved shear stress, shear strain data of Fig. 4.6 were calculated for single slip by means of Eqs. (4.12) and (4.13). The initial crystallographic orientations for these experiments and for the 99.47% purity experiments of Figs. 4.7, 4.8, and 4.10 were shown in Fig. 4.5.

A second series of dead-weight tension uniaxial stress experiments were carried out by the writer on 99.47% aluminum. These single crystal specimens had a diameter of 1.24 mm with lengths which varied from 30 mm to 110 mm for an average length-to-diameter ratio of 62. The crystals were grown by the strain anneal method. The purpose of the experiments of this series was twofold. The major purpose was to obtain dead-weight data with extreme accuracy, both in axial stress and strain

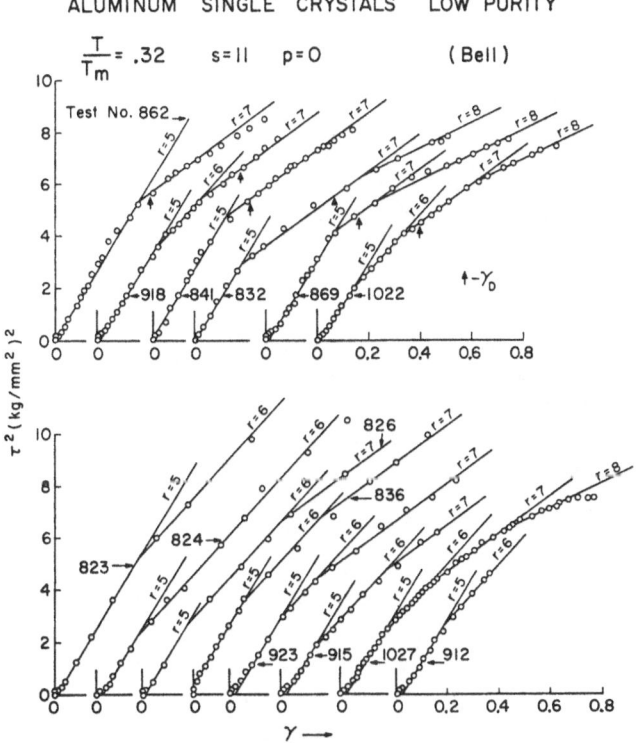

Fig. 4.7. τ^2 vs γ plots of medium purity aluminum single crystals resolved in macroscopic single slip. Avg. first $\gamma_N = 23^0/_0$ $(r = 5)$

measurement and in the x-ray diffraction determination of crystallographic orientation. Such data not only would allow a comparison of single crystal and polycrystalline behavior in terms of the parabolic generalization, but also would serve as a norm with respect to evaluating the accuracy of such measurements in general.

A second reason for performing these experiments was to provide a test of the double slip deformation hypothesis. For that purpose, x-ray diffraction measurements were made at a large number of repeated intervals during the experiments.

The apparatus for accomplishing these objectives is shown in Fig. 4.9. The load was dead-weight and the strain was determined by an optical cathetometer. A clamping device was provided which permitted x-ray diffraction measurements to be made while the specimens were under load.

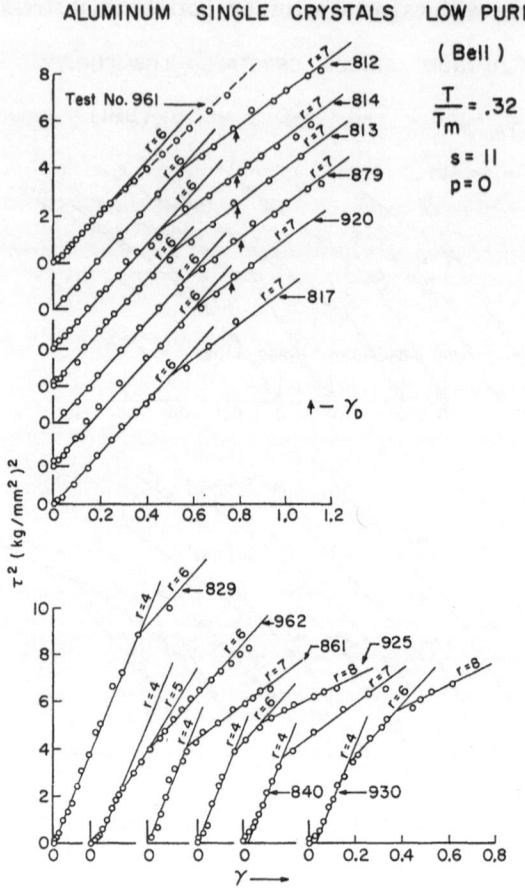

Fig. 4.8. τ^2 vs γ plots of medium purity aluminum single crystals resolved in macroscopic single slip. Avg. first $\gamma_N = 50.5\%$ ($r = 6$); 23% ($r = 4$)

The polycrystalline 99.45% purity quasi-static uniaxial stress experiments of Fig. 3.5 provided initial parabolas with mode indices of $r = 4$, $r = 5$, and $r = 6$. This is precisely the situation for the 99.47% aluminum single crystal data shown in Figs. 4.7 and 4.8. Averaged transitions also occur at predicted $\gamma_N = 23\%$ or $\gamma_N = 50.5\%$. Fourteen of the 27 single crystal experiments had an initial parabola with a mode index of $r = 5$, 6 had a value $r = 4$, and 7 had a value of $r = 6$. One first observes that the experimental data (circles) is correlated with predicted parabolas (solid lines) from Eq. (4.11). Nine of the 14 experiments with an initial mode index $r = 5$ had unit transitions rather than the double transitions expected for tension tests. This also was true of all 7 of the experiments with an initial mode index of $r = 6$. All but one of the experiments with initial mode index of $r = 4$ had the characteristic

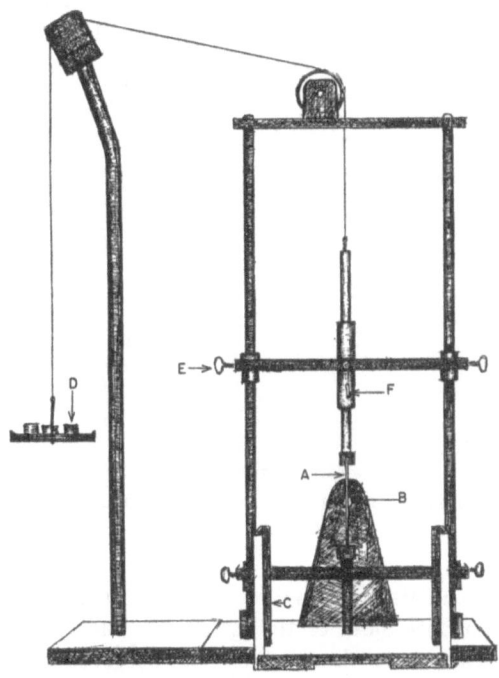

Fig. 4.9. Tension testing apparatus: A. specimen; B. x-ray collimator; C. x-ray film cassette holder; D. loading weights; E. adjusting clamp for axiality guide; F. specimen clamp for x-ray analysis

double or triple second-order transition of polycrystalline tension experiments. Aluminum of 99.47% purity not only has a single crystal behavior in close agreement with the writer's parabolic large deformation generalization, but also has an initial mode index distribution in quantitative agreement with polycrystalline experiments on aluminum of the same purity. The experiments of Figs. 4.3 and 4.4 which were used by BELL and GREEN (1967) in the study of the double slip deformation hypothesis were part of the 99.47% purity experiments of Figs. 4.7 and 4.8, as a comparison of test designations will reveal.

Specimens 829, 832, 836, and 862 were annealed at 620° C for 23 hours and furnace cooled; all others were tested as grown. Fourteen of these experiments rotated to, and beyond the $\langle 100 \rangle - \langle 111 \rangle$ symmetry line.

All of the data of Figs. 4.6, 4.7, 4.8, and 4.10 were calculated for single slip. The deformation proceeds along the same parabola across the symmetry line in every instance, as may be seen from the values of γ_D in Figs. 4.7 and 4.8 which indicate the strain at which the deformation reached the symmetry line. This strain is designated by arrows. In the BELL and GREEN (1967) study, double slip calculations and single slip calculations were compared (Fig. 4.4) with resolved shear stress, resolved

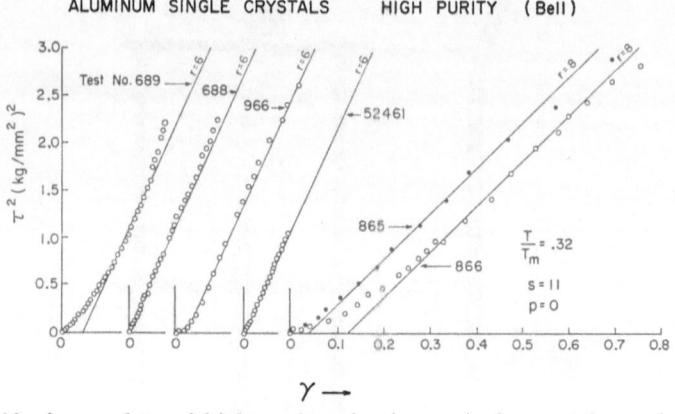

Fig. 4.10. τ^2 vs γ plots of high purity aluminum single crystals resolved in macroscopic single slip

shear strain data calculated from direct x-ray measurements for specimen axis rotations on both sides of the symmetry line. It was shown from those direct measurements that the behavior observed in Figs. 4.7 and 4.8 in terms of single slip calculations, was that which pertained.

The third group of uniaxial stress, single crystal experiments were carried out in $99.99 +\%$ purity aluminum. Experiments Nos. 688, 689, and 52461 were uniaxial compression measurements on 1 in. diameter 3 in. long specimens. Tests 865, 866, and 966 were for dead-weight uniaxial tension experiments in 1.24 mm diameter specimens. All of these experiments, as shown in the τ^2 vs γ plots of Fig. 4.10, provided parabolic stress-strain functions with mode indices in agreement with prediction from the generalized large deformation behavior of Eq. (4.11). The absence of transitions in the deformation of such high purity specimens obviously is of importance in understanding this phenomenon.

The single crystal parabola mode index of $r = 6$ for the experiments of Fig. 4.10 is precisely the same as that for 10 of the polycrystalline measurements for 99.99% purity aluminum shown in Fig. 3.5. For both polycrystals and single crystals, the value of $r = 6$ occurs in both tension and compression experiments. One polycrystalline dead-weight compression experiment in 99.99% purity aluminum at 293° K had an initial deformation mode index of $r = 8$ similar to that in the two dead-weight tension single crystal experiments of Fig. 4.10. Several dead-weight compression polycrystalline experiments, whose data were shown in Figs. 3.5, 3.6, and 3.7 of Chapter III, at temperatures above room temperature have a mode index of $r = 8$ for 99.99% polycrystalline aluminum. The single crystal tests 865 and 866 possess an initial stage I and stage II deformation so that straight-line parabolicity of the stage III in a τ^2 vs γ plot does not begin until the stress τ^* is reached (Fig. 4.1).

A series of diffraction grating finite amplitude compression wave propagation experiments carried out in the writer's laboratory by GILLICH (1964, 1967) have provided parabola coefficients in 99.99% aluminum single crystals with deformation mode indices of $r = 6$ and $r = 7$. The quasi-static uniaxial compression experiments, Nos. 688 and 689 of Fig. 4.10, were performed by GILLICH for the purpose of comparison with his dynamic compression single crystal data. The quantitative correlation of quasi-static single crystal parabolas and finite amplitude wave propagation data once again is demonstrated in these comparisons.

GILLICH obtained a second set of wave propagation data from which he proposed that there could exist in some instances a fourth power law in addition to the more common parabolic behavior. An alternative interpretation of these same dynamic single crystal GILLICH data is that the differences in initial orientation of the projectile specimen and the struck specimen produced variations in first diameter wave initiation, with a corresponding distortion of strain arrival times outside the first diameter and changes in the mode index r.

Because of the difficulty of obtaining a symmetrical impact of identical specimens when two large single crystals are in axial collision, GILLICH was unable to relate maximum strains to impact velocities and thus meet the particle velocity vs strain condition, Eq. (2.4), of the finite amplitude wave theory. In polycrystalline α-brass, HARTMAN (1967) also has observed in a few instances low strain arrival times consistent with a fourth power law. However, symmetrical impacts were obtained so that maximum strain and impact velocities could be correlated. HARTMAN's data demonstrates that the deformation is governed by a parabolic law with a different mode index r. This is the behavior, suggested above, for the single crystal dynamic data with a deformation mode of $r = 5$. The differences in arrival times on the few occasions when such differences are present, may be a first diameter phenomenon, probably arising from slight non-axiality in the collision.

In Fig. 4.11 are shown strain-time diffraction grating measurements from GILLICH's (1964, 1967) finite amplitude wave propagation study in

99.99% aluminum single crystals. Both types of tests are compared with predicted arrival times.

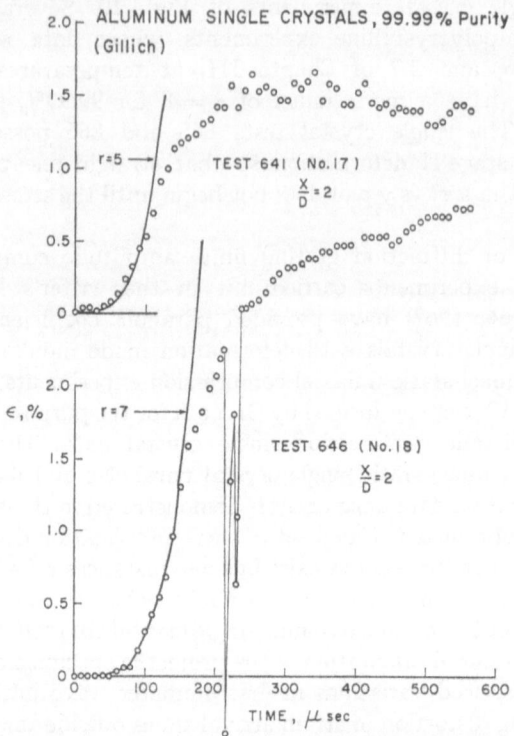

Fig. 4.11. Two high purity aluminum single crystal diffraction grating measurements (circles) compared with parabolic stress-strain prediction (solid lines) from Eq. (4.11) for $r = 5$ and $r = 2$. Arrival times of test 17 also provide a $^1/_4$ power law (GILLICH, 1964). Note spikes in the data produced by the sudden increase in light on photomultiplier tubes when new deformation bands are exposed

The comparison of the approximately 50 quasi-static polycrystalline experiments of Figs. 3.3 to 3.7 with the same number of single crystal experiments of Figs. 4.6, 4.7, 4.8, and 4.11 provides unquestionable experimental evidence that the large deformation generalization of Eqs. (3.1) and (4.11) is applicable for all three purities investigated. If one includes the finite amplitude wave propagation experiments in aluminum, from which the generalization was first discovered, then it may be stated as an experimental fact that the writer's parabolic generalization with its deformation mode and transition structure is applicable for strain rates from $\dot{\varepsilon} = 10^{-5}$ sec^{-1} to $\dot{\varepsilon} = 10^4$ sec^{-1}, a ratio of strain rates of 10^9.

In the further exploration of the applicability of the generalization at extremely low strain rates, two uniaxial stress experiments in quasi-static compression are shown in Figs. 4.12 and 4.13. The polycrystalline experiment was a dead-weight test which was continually monitored for five days. A 0.0067 kg/mm² dead-weight load was added on an average of every 30 minutes during the 5-day interval.

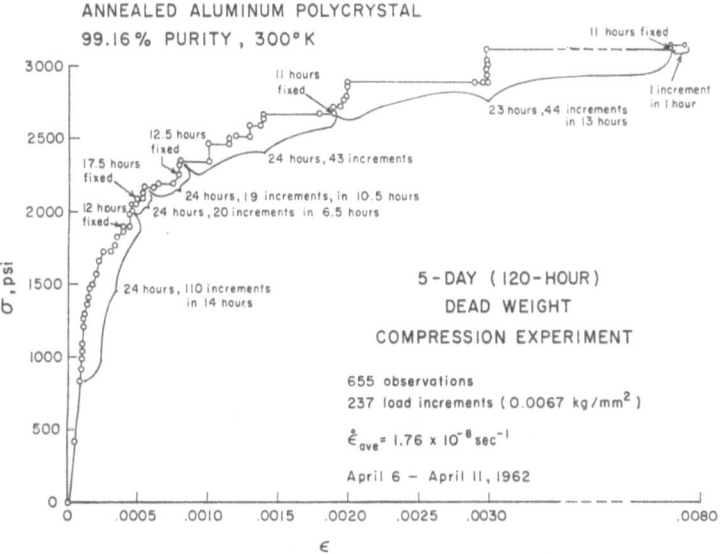

Fig. 4.12. A continuously monitored incremental compression experiment exhibiting the Portevin - le Chatelier effect in polycrystalline aluminum. $\dot{\varepsilon}_{avg} = 1.76 \times 10^{-8}$ sec⁻¹

The second experiment shown in Fig. 4.13 is a single crystal measurement for which the total time during which the experiment was monitored was fifteen days; the load increment was also 0.0067 kg/mm². The average strain rate for the polycrystalline test was $\dot{\varepsilon} = 10^{-8}$ sec⁻¹ and for the single crystal test, the extremely low average strain rate of $\dot{\varepsilon} = 10^{-9}$ sec⁻¹. Both specimens were 1 in. in diameter and 3 in. long. The polycrystalline specimen was 99.16% pure, while the single crystal purity was 99.99%. Both of these experiments will be referred to in a later chapter which considers the importance of the Portevin - le Chatelier effect with respect to the mode and transition structure. Both experiments of Figs. 4.12 and 4.13 provided parabolas in agreement with the generalization even at the extremely low strain rates.

A comparison of the data of Figs. 4.12 and 4.13 with wave propagation experiments extends the ratio of strain rates to which the present

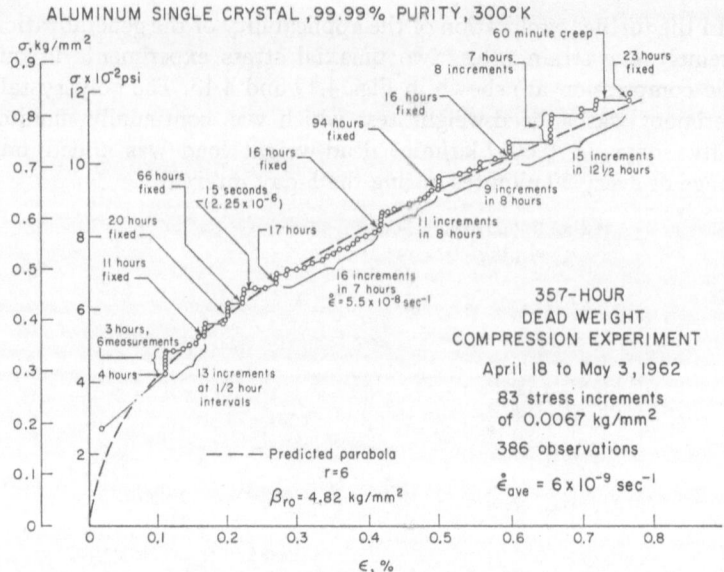

Fig. 4.13. A continuously monitored incremental compression experiment exhibiting the Portevin - le Chatelier effect at $\dot{\varepsilon} = 6 \times 10^{-9}\ \text{sec}^{-1}$ in a high purity aluminum single crystal

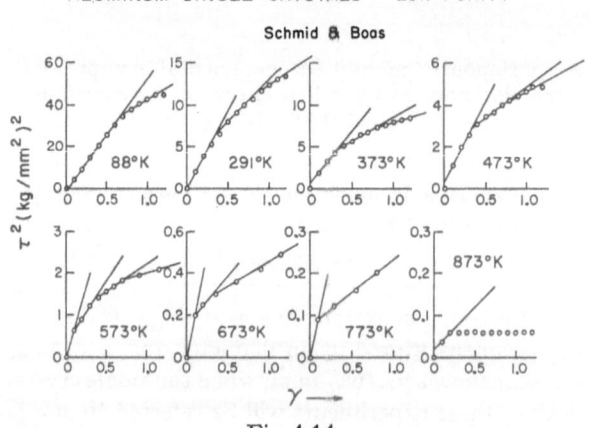

Fig. 4.14

large deformation generalization applies, to 10^{13}. This is an experimental observation not based upon theoretical hypothesis. Any theoretical treatment of the resolved stage III large distortional deformation behavior in the element, aluminum, must be concerned with the stability properties of the solid under conditions prescribing a zero viscous contribution.

In (BELL, 1964, 1965 a) the writer compared the calculated $\beta_{ro}(II)$ and $\beta_{ro}(III)$ zero-point parabola coefficients of nearly 400 single crystal

ALUMINUM SINGLE CRYSTALS   HIGH PURITY

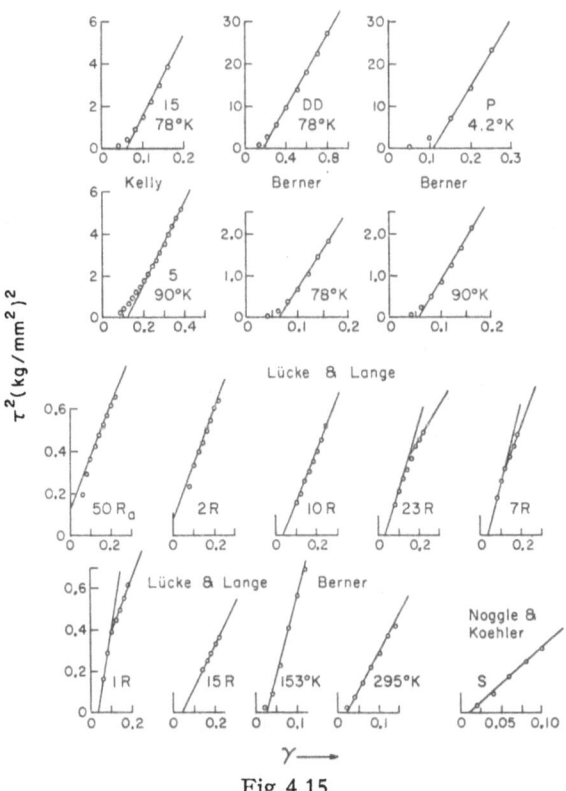

Fig. 4.15

measurements in the literature with a somewhat different form of the present large deformation generalization. The data of the designated experimentalists are listed in Appendix I together with the observed stage II slopes, θ_{II}, and stage II—stage III stress, τ^*. The τ^2 vs γ plots of the stage III deformation of those aluminum experiments from the literature are shown in Figs. 4.14, 4.15, 4.16, and 4.17. From these literature data, as was shown above for the writer's 50 single crystal experiments in aluminum, a general agreement with prediction is obtained, irrespective of the differences in purity, strain rates, or specimen dimensions, and irrespective of the differences in ambient temperature.

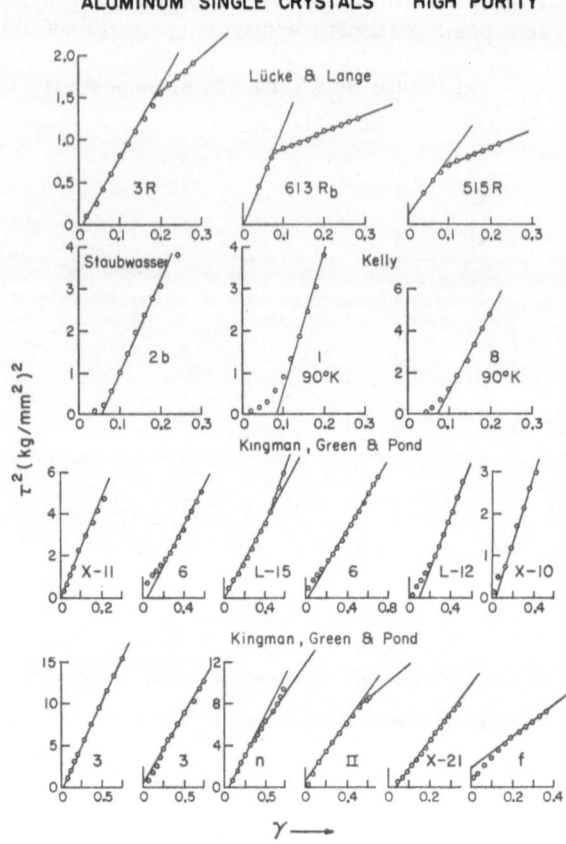

Fig. 4.16. τ^2 vs γ plots in high purity aluminum in initial corner orientations

Several years ago, in examining the very well known early aluminum experiments of BOAS and SCHMID (1931), the writer first observed in single crystals the transition structure which characterized the finite amplitude wave initiation behavior of polycrystals. These low purity data of BOAS and SCHMID (1931) are shown in the τ^2 vs γ plots of Fig. 4.14. The initial parabola coefficient of $r = 5$ and the subsequent transitions for these data of 35 years ago are characteristically the same as the recent data of the present writer in the same purity aluminum shown in Figs. 4.7 and 4.8.

As was pointed out in the introduction to the present chapter, single crystal data expressed in terms of a resolved shear stress, resolved shear strain function are subject to an additional source of error in the measurement of x-ray diffraction angles. From Eqs. (4.2), (4.12), and (4.13) this error for a parabolic deformation may be determined from Eq. (4.17).

$$\beta^2 = \frac{\tau^2}{\gamma} = \sigma^2 \frac{\cos^3 \varphi_0}{(1+\varepsilon)^2} \frac{(1+\varepsilon)^2 - \sin^2 \lambda_0}{[(1+\varepsilon)^2 - \sin^2 \lambda_0]^{1/2} - \cos \lambda_0} . \qquad (4.17)$$

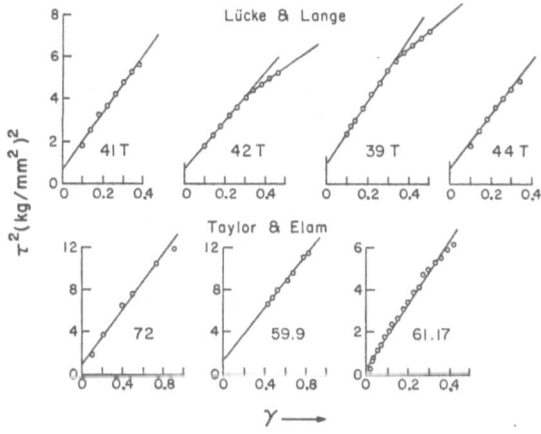

ALUMINUM SINGLE CRYSTALS LOW PURITY

Fig. 4.17

The magnitude of the error in the parabola coefficient which would arise from the inaccurate x-ray diffraction determination of φ_0 and λ_0 would depend not only upon the magnitude of those values but also upon the magnitude of the axial stress σ and axial strain ε. For a SCHMID factor of 0.5 with $\lambda_0 = \varphi_0 = \frac{\pi}{4}$, with a stress $\sigma = 10$ kg/mm² and $\varepsilon = 0.1$, a simultaneous plus and minus 1° error in φ_0 and λ_0 produces an error in β of approximately 5%. Thus, instead of the predicted parabola coefficient of $\beta = 10.85$ kg/mm² one would obtain a parabola coefficient of 10.31 kg/mm² or 11.39 kg/mm² depending upon which of the measured angles had a positive 1° error, and which had a negative 1° error. A simultaneous 1° error of the same sign produces an error in β of not quite 1%. The situation chosen for illustration represents a maximum value since the SCHMID factor $\cos \varphi_0 \cos \lambda_0$ of 0.5 was assumed.

In one investigation (BELL, 1964, 1965a) stage III single crystal parabola coefficients in 6 different face-centered cubic solids were introduced into the TAYLOR (1938) aggregate theory to determine the zero-point parabola coefficient of low purity completely annealed polycrystalline aluminum. The data of over 300 single crystal experiments when averaged, provided a close agreement between calculated and experimentally determined polycrystalline parabola coefficients. To perform this calculation, an empirical factor $\left(\frac{3}{2}\right)^{3n/2}$ was introduced where n assumed one of the integral values — 1, 0, 1, 2, or 3 (BELL, 1965a).

As will be seen in Chapter V, the specific form of the large deformation generalization of Eqs. (1.1) and (4.11) evolved from the study of this empirical factor $\left(\frac{3}{2}\right)^{3n/2}$. In the present chapter the stage III single crystal data (BELL, 1965a) are averaged for a comparison with the large deformation generalization in its present form.

In Appendix I are tabulated $\beta_{ro}(II)$, $\beta_{ro}(III)$, and θ_{II} data for 375 single crystal experiments in Al, Ag, Au, Ni, Pb, Cu, Fe, and Ta. For each zero-point parabola coefficient the nearest value of the mode index r is given. The tabulated data of Appendix I, of course, already include the temperature dependence factor $(1 - T/T_m)$. In Table I are

Table I. *Averaged Al data from Appendix I*

Mode Index r	No. of Measurements	Average $\beta_{ro}(II)$ Experiments	Theoretical $\beta_{ro}(II)$	No. of Measurements	Average $\beta_{ro}(III)$ Experiments	Theoretical $\beta_{ro}(III)$
1	1	13.38	13.28	1	12.78	13.28
2	—	—	—	1	10.15	10.84
3	—	—	—·	8	8.05	8.85
4	—	—	—	23	6.98	7.23
5	6	5.94	5.90	76	5.86	5.90
6	—	—	—	52	4.77	4.82
7	4	4.15	3.94	42	4.02	3.94
8	3	3.28	3.21	30	3.17	3.21
9	2	2.49	2.62	22	2.705	2.62
10	1	2.30	2.14	24	2.20	2.14
11	—	—	—	10	1.77	1.75

shown averaged $\beta_{ro}(II)$ and $\beta_{ro}(III)$ aluminum stage III single crystal data for each mode index r. These data are compared with predicted single crystal parabola coefficients β_{ro} in terms of the present large deformation generalization; i.e., $\beta_{ro} = \left(\frac{2}{3}\right)^{r/2} \frac{\mu(0)\,B_0}{m^{3/2}}$. The data in Table I do not include this writer's 50 aluminum single crystal experiments which have been discussed separately in the present chapter. τ^2 vs γ plots of these same experiments for high purity single crystals, and in Fig. 4.17, low purity single crystals are shown. Additional low purity data of BOAS and SCHMID (1931) also tabulated in Appendix I, were shown in Fig. 4.14.

As may be seen from these τ^2 vs γ plots, stage III deformation is parabolic. Averaged experimental parabola coefficients for the stage III deformation of aluminum from $T/T_m = 0.005$ to $T/T_m = 0.937$ are consistent with the large deformation generalization. The writer's aluminum single crystal data of Figs. 4.6 through 4.10 were for dead-weight loading. The single crystal data of Table I representing over 40 years of experimentation, were performed under a wide variety of differing loading histories. Nevertheless, one sees that a general correlation has been obtained.

A series of experiments has been carried out in the writer's laboratory in which the large deformation behavior of more than one single crystal grain has been determined during the same experiment. SHARPE (1966a, b, c), in his study of the effect of grain boundaries on the Porte-vin-le Chatelier effect, conducted two such experiments for constant stress rate. Measurements were made on bi-crystals in which in one instance the single grain boundary was approximately perpendicular to the cylindrical axis; i.e., a bamboo crystal (Fig. 4.18).

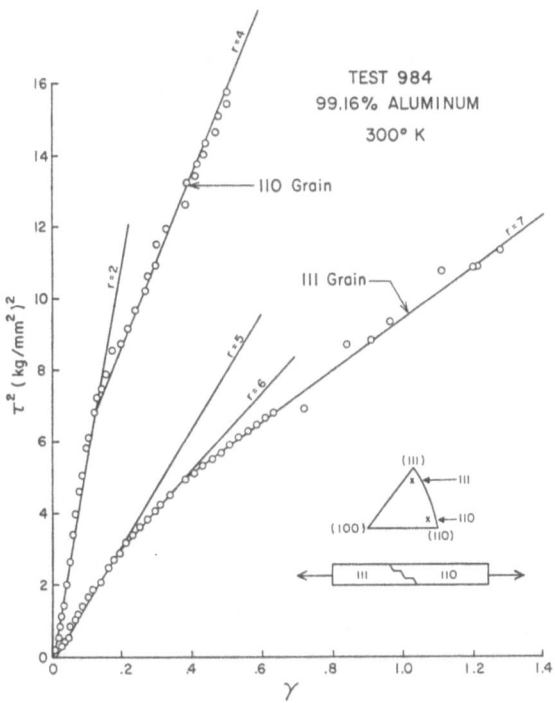

Fig. 4.18. Simultaneous deformation behavior of each member of a bamboo bi-crystal, showing different deformation modes in each crystal

A second measurement was made in which the single boundary was approximately parallel to the cylindrical axis (Fig. 4.19).

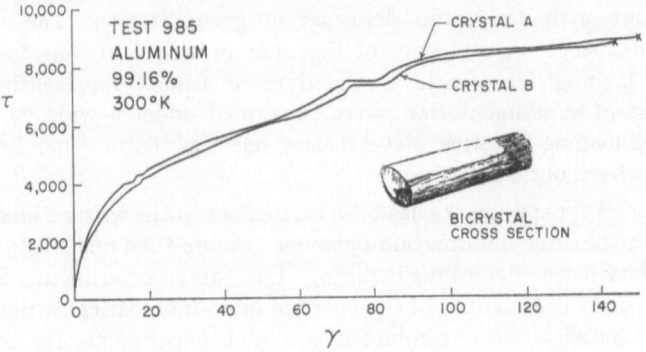

Fig. 4.19. Simultaneous measurement of resolved deformation in each member of a longitudinal bi-crystal, showing common behavior

In each instance strain measurements were made on the individual crystals of the bi-crystal. X-ray diffraction measurements were made on each grain so that individual resolutions could be made. For the bamboo crystal the data for the $\langle 110 \rangle$ grain already has been shown in Fig. 4.6 as test No. 984. This measurement provided a τ^2 vs γ plot identical with that of all the other single crystals of the same purity; i.e., initial mode index $r = 2$ with a transition to $r = 4$ at the second transition strain of $\gamma_N = 0.13$.

The $\langle 111 \rangle$ grain, however, as shown in Fig. 4.18 has an initial parabola coefficient mode index of $r = 5$, followed by a unit transition to $r = 6$ in the vicinity of the third transition strain $\gamma_N = 0.23$, and a second unit transition to $r = 7$. Failure occurred on this $\langle 111 \rangle$ grain. One notes that contrary to expectations the $\langle 111 \rangle$ grain lies below the $\langle 110 \rangle$ grain. Single slip resolution was used for both experiments.

The bi-crystal experiment of Fig. 4.19 also was included in Fig. 4.6 as test No. 985. In this instance, however, as may be seen from Fig. 4.19, the resolved τ vs γ data is nearly identical for each grain of the longitudinal bi-crystal. A third experiment was performed in dead-weight uniaxial compression for an aluminum multi-crystal containing a single large grain in a small grain matrix. Strain was measured separately on the large grain and on the multi-crystal matrix by means of an optical cathetometer. The specimen and the measured deformation are illustrated in Fig. 4.20.

The experimental data obtained for the present writer by C. JEFFUS shows that the multi-crystal portion of the data closely follows the

$r = 6$ parabola characteristic of experiments in 99.99% purity poly-crystals as shown in Fig. 3.5. The σ—ε behavior of the individual grain based upon predicted resolved stage III and deresolved macroscopic single slip also has an $r = 6$ parabola.

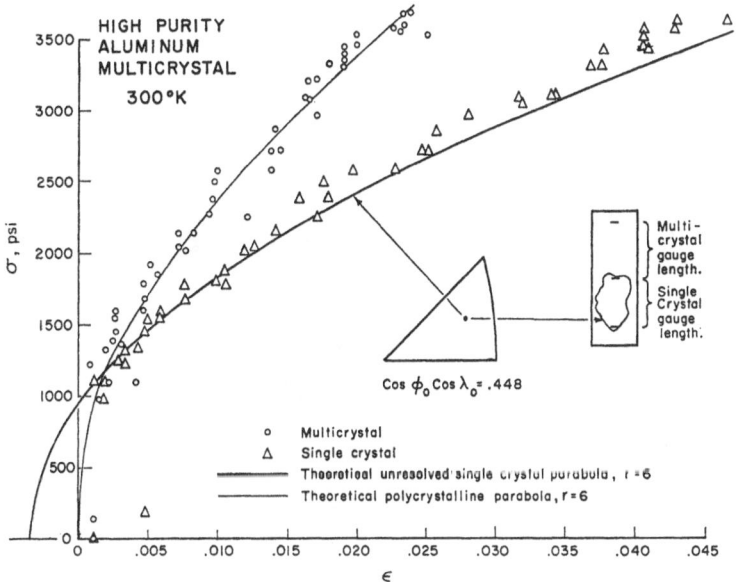

Fig. 4.20. Experimental study of a large grain in an otherwise medium-grain high purity multicrystal. The multicrystal is compared with Eq. (3.1) for $r = 6$, and the single crystal, which has been unresolved in macroscopic single slip, is compared with Eq. (4.11)

Thus we see from this examination of the behavior of single crystals deforming in the presence of adjacent crystals, that individual grains deform simultaneously in the same deformation mode despite alternation of the Portevin - le Chatelier effect between one grain and another.

Gold

The crystalline element gold, like the element silver, is of special interest for a comparison with aluminum. The shear moduli of these three elements are nearly identical as are the atomic lattice spacings. The melting points T_m and mass densities ϱ are, of course, widely different. Stage III gold single crystal data experiments performed over a span of 34 years are compared in Fig. 4.21 with prediction from Eq. (4.11) for a mode index $r = 5$. The value of γ_b was arbitrarily chosen.

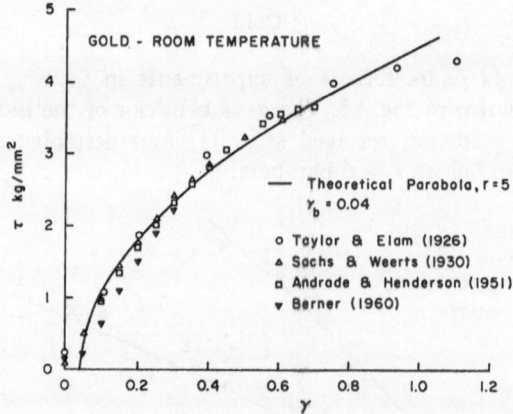

Fig. 4.21. A comparison of resolved gold single crystal data obtained during the past 40 years. The solid line is from Eq. (4.11) for $r = 5$

GOLD SINGLE CRYSTALS

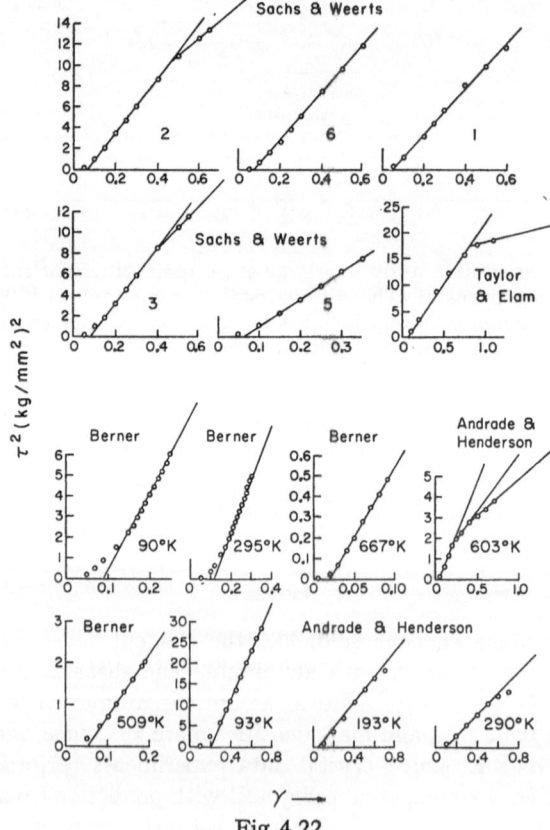

Fig. 4.22

The melting point T_m of gold is 1,334° K while that for aluminum is $T_m = 932°$ K. The room temperature parabola (solid line) of Fig. 4.21 for gold differs from the room temperature data in aluminum solely by the difference in fractional melting point temperatures. Thus, the solid line in Fig. 4.21 may be considered as a prediction of the large deformation room temperature behavior of gold, from aluminum. These same data, plus additional data at other temperatures from 90° K to 667° K, are shown in τ^2 vs γ plots in Fig. 4.22. The circles in each instance represent the experimental points.

The solid line is the best straight line drawn through these data for the purpose of calculating the value of $\beta_{ro}(III)$ tabulated in Appendix I.

An inspection of Fig. 4.22 reveals that the gold stage III distortional deformation is unquestionably parabolic. The $\beta_{ro}(II)$ zero-point parabola coefficients determined from Eq. (4.5) and the $\beta_{ro}(III)$ parabola coefficients determined from Eq. (4.7) tabulated in Appendix I provide averaged parabola coefficients for the 14 gold single crystal measurements compared with prediction in Table II.

Table II. *Averaged Au data from Appendix I*

Mode Index γ	No. of Measurements	Average $\beta_{ro}(II)$ Experiments	Theoretical $\beta_{ro}(II)$	No. of Measurements	Average $\beta_{ro}(III)$ Experiments	Theoretical $\beta_{ro}(III)$
3	—	—	—	1	8.32	8.85
4	2	6.775	7.23	5	6.63	7.23
5	7	6.04	5.90	8	6.12	5.90
6	3	5.21	4.82	2	5.22	4.82
7	—	—	—	1	4.28	3.94
8	—	—	—	1	3.46	3.21
9	—	—	—	1	2.62	2.62

Silver

A series of 9 single crystal experiments of ANDRADE and HENDERSON (1951) in silver are shown in Fig. 4.23.

The temperatures at which these experiments were performed varied from 93° K to 623° K. The stage II calculations of a total of 16 silver tests in this temperature range have been included in Appendix I, of which only the 9 τ^2 vs γ plots of Fig. 4.23 are shown here.

As may be seen from the data of Fig. 4.23, high purity silver is more subject to second-order transitions than is high purity gold. This is further demonstrated from an examination of the initial parabola coefficients for silver in Appendix I where mode indices of $\gamma = 3$, $\gamma = 4$, and $\gamma = 5$ may be observed for both $\beta_{ro}(II)$ and $\beta_{ro}(III)$. These experimental parabola coefficients are compared with prediction in Table III.

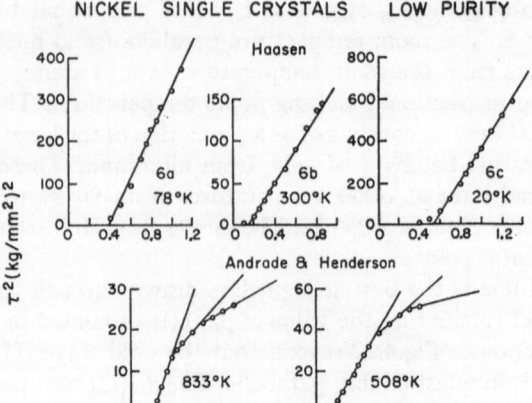

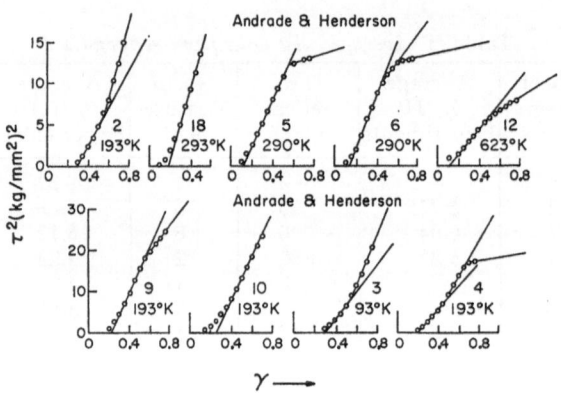

Fig. 4.23

Table III. *Averaged Ag data from Appendix I*

Mode Index γ	No. of Measurements	Average $\beta_{ro}(II)$ Experiments	Theoretical $\beta_{ro}(II)$	No. of Measurements	Average $\beta_{ro}(III)$ Experiments	Theoretical $\beta_{ro}(III)$
3	3	8.49	8.85	4	8.59	8.85
4	6	7.17	7.23	5	7.46	7.23
5	6	5.78	5.90	4	6.13	5.90
6	—	—	—	2	5.25	4.82
9	—	—	—	1	2.60	2.62
10	—	—	—	2	2.22	2.14

It may be seen that silver not only has a stage III deformation which is parabolic, but also has experimental parabola coefficients in agreement with the writer's large deformation generalization.

Nickel

The total number of nickel single crystal stage III deformation experiments tabulated in Appendix I is 35 with 15 low purity measurements and 20 high purity measurements. These experiments have been performed at temperatures ranging from 20° K to 833° K. Twenty-three of the experiments are from the work of MADER, SEEGER, and LEITZ (1963). In these 20 high purity single crystals and 3 low purity single crystals the data were given only in the form of tabulated θ_{II} and τ^*. However, these two stage II parameters are sufficient in terms of Eq. (4.5) to determine $\beta_{ro}(II)$ zero-point parabola coefficients.

The τ^2 vs γ plots of Fig. 4.23 for the low purity single crystal data of HAASEN (1958) and of ANDRADE and HENDERSON (1951) demonstrate that the stress-strain function possesses the requisite parabolicity. The data of the 15 low purity nickel experiments have a different distribution of mode indices than do the high purity data. The experimental and predicted parabola coefficients for these mode indices are compared in Table IV.

Table IV. *Averaged Ni data from Appendix I*

Mode Index ν	No. of Measurements	Average $\beta_{ro}(II)$ Experiments	Theoretical $\beta_{ro}(II)$	No. of Measurements	Average $\beta_{ro}(III)$ Experiments	Theoretical $\beta_{ro}(III)$
2	2	30.455	29.92	2	27.72	29.92
3	—	—	—	1	24.20	24.43
4	7	19.81	19.94	—	—	—
5	3	15.89	16.28	2	16.555	16.28
6	7	12.88	13.30	1	13.09	13.30
7	13	11.03	10.86	1	10.13	10.86
8	4	9.51	8.86	1	9.20	8.86
9	—	—	—	1	7.29	7.24
11	—	—	—	1	4.91	4.83

As was characteristic of nickel polycrystalline data, whether dynamic or quasi-static, it may be seen that stage III deformation of nickel single crystals, whether of high or low purity, is expressible in terms of the linearly temperature dependent generalized parabolic large deformation stress-strain function, Eq. (4.11).

Lead

The lead single crystal data which have been included in this study are those of BOLLING, HAYS, and WIEDERSICH (1962). In their presentation of experiments no actual test data were included; instead, tabulated values of θ_{II} and τ^* were provided from which $\beta_{ro}(II)$ parabola coefficients could be determined by means of Eq. (4.5). The 28 experiments which BOLLING et al. (1962) conducted are tabulated in Appendix I for high purity lead which were performed at temperatures from 4.2° K to 273° K. Experimental parabola coefficients are compared with prediction in Table V.

Table V. *Averaged Pb data from Appendix I*

Mode Index r	No. of Measurements	Average $\beta_{ro}(II)$ Experiments	Theoretical $\beta_{ro}(II)$	No. of Measurements	Average $\beta_{ro}(III)$ Experiments	Theoretical $\beta_{ro}(III)$
1	4	3.205	3.23	—	—	—
2	1	2.82	2.64	—	—	—
3	3	2.21	2.155	—	—	—
4	8	1.71	1.76	—	—	—
5	11	1.45	1.44	—	—	—
6	1	1.25	1.17	—	—	—

These data provide a sufficiently close correlation with prediction so that stage III deformation of high purity lead single crystals may be included within the framework of the present generalization. It is interesting that the quasi-static and dynamic finite amplitude wave experiments of SPERRAZZA (1961, 1962a, b) discussed in Chapters II and III provided a polycrystalline high parabola coefficient for lead of $r = 4$ at room temperature, which is consistent with the range of single crystal mode indices in Table V at room temperature. The BOLLING et al. (1962) data at 77° K were performed at strain rates varying from 10^{-2} to 10^{-6} sec^{-1}. That the corresponding parabola coefficients for finite distortional deformation are independent of strain rate even in quasi-static deformation may be seen from an inspection of the data of Table VI.

Table VI. *High purity Pb (Bolling et al.)*

$\dot{\gamma}$ sec^{-1}	Number of tests	$\beta_{ro}(II)$ (kg/mm^2)
10^{-2}	3	1.65
10^{-3}	2	1.48
10^{-4}	15	1.67
10^{-5}	2	1.54
10^{-6}	1	1.55

Copper

In examining the single crystal deformation literature, one finds that copper is second only to aluminum in the number of experiments which have been reported. This writer has shown that for high purity copper experiments stage II and stage III deformation are interrelated in the manner prescribed by Eqs. (4.5), (4.6), and (4.7) (BELL, 1964, 1965a). It was also shown that the stage III deformation parabolicity was independent of the length of the stage I easy glide, or of the slope θ_{II} of stage II (BELL, 1964). This independence of stage III behavior from earlier deformation may be seen in Fig. 4.2 where the data of the designated experimentalists (solid lines) are compared with prediction (circles) obtained from introducing their experimental values θ_{II}, τ^*, and γ^* into Eqs. (4.5) and (4.6).

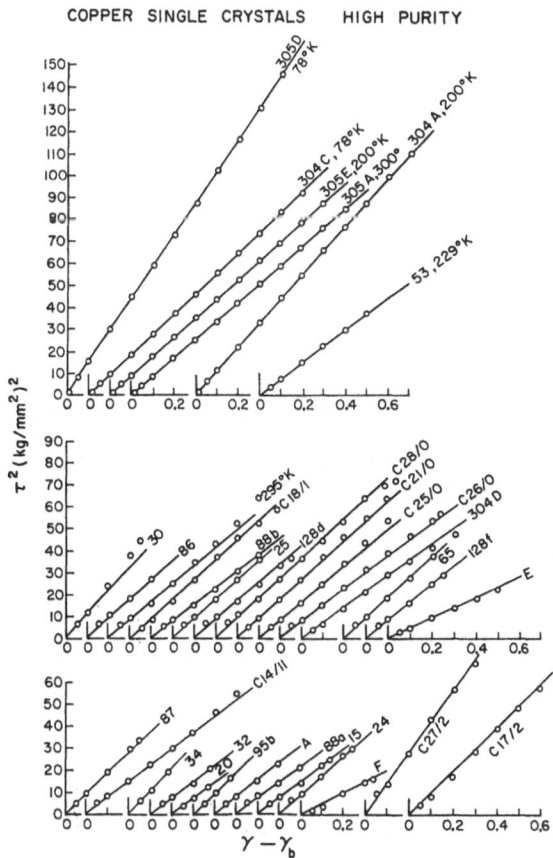

Fig. 4.24. τ^2 vs $(\gamma - \gamma_b)$ plots for high purity copper single crystals. Unless otherwise marked, experiments are at room temperature

In Figs. 4.24 and 4.25 are shown the τ^2 vs $(\gamma - \gamma_b)$ and τ^2 vs γ plots of 40 experiments at temperatures ranging from 78° K to 1173° K. That straight lines may be drawn through all of these data demonstrates the general parabolicity of the stage III copper single crystals, whether of high or low purity. Unless designated otherwise, the data of Fig. 4.24 are at room temperature.

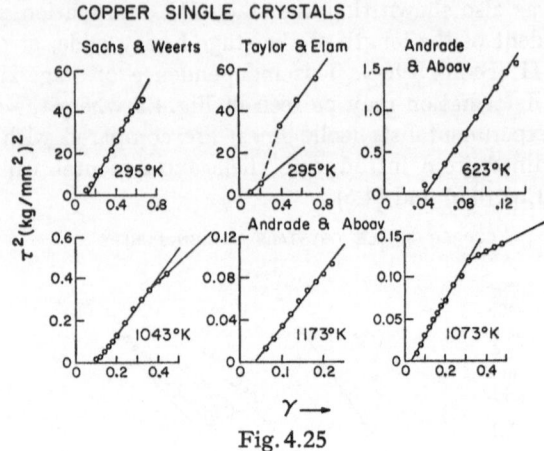

Fig. 4.25

A comparison of the $\beta_{ro}(II)$ and $\beta_{ro}(III)$ experimental data with prediction is shown in Table VII.

Table VII. *Averaged Cu data from Appendix I*

Mode Index γ	No. of Measurements	Average $\beta_{ro}(II)$ Experiments	Theoretical $\beta_{ro}(II)$	No. of Measurements	Average $\beta_{ro}(III)$ Experiments	Theoretical $\beta_{ro}(III)$
1	4	21.82	22.03	—	—	—
2	11	18.10	17.99	3	17.56	17.99
3	17	14.88	14.69	7	15.56	14.69
4	23	11.71	11.99	34	11.83	11.99
5	22	10.11	9.79	6	9.93	9.79
6	6	8.53	8.00	2	8.03	8.00
8	2	5.225	5.33	2	5.22	5.33
11	—	—	—	1	3.09	2.90
12	—	—	—	2	2.30	2.37

The experimental data of MITCHELL and THORNTON (1963) which have been included in Appendix I were not given in the form of experimental stress-strain functions but as tabulated values of θ_{II} and τ^* from which, by means of Eq. (4.5), the listed $\beta_{ro}(II)$ parabola coefficients were determined.

There are many other aspects of the copper data which might be explored here. Among these are the experimental data of BERNER (1957) in high purity copper at room temperature which were carried out for strain rates from $\dot{\varepsilon} = 4.46 \times 10^{-6}$ sec^{-1} to $\dot{\varepsilon} = 1.27 \times 10^{-2}$ sec^{-1}. BERNER's data are shown in Fig. 4.26 and are also tabulated in Appendix I.

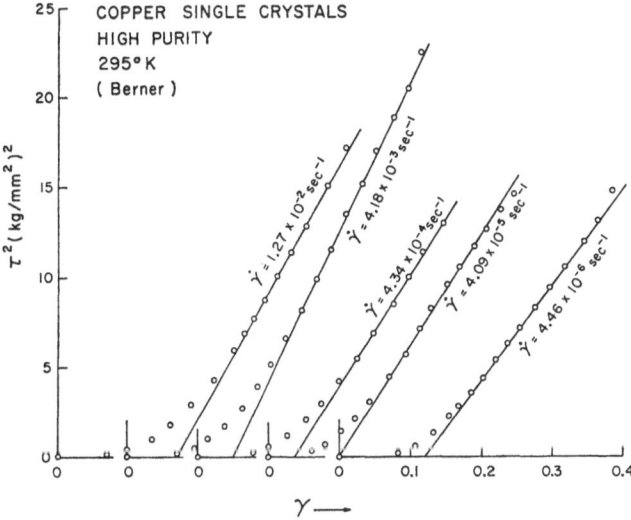

Fig. 4.26. τ^2 vs γ plots for stage III high purity copper single crystals at 295° K. Note lack of dependence upon strain rate over 4 orders of magnitude

Not only is the stage III portion obviously parabolic at all strain rates, but the variation of strain rates by 4 orders of magnitude at a single ambient temperature provides no evidence that viscosity is of importance in the quasi-static resolved stress-strain function of high purity copper. It already has been shown, of course, in the wave propagation studies described in Chapter II and in the quasi-static experiments described in Chapter III that the finite deformation of polycrystalline copper is independent of viscosity for variations in strain rates of 8 orders of magnitude.

Iron

All of the single crystals discussed thus far in this chapter have had a face-centered cubic crystalline structure. Both iron and tantalum, for which stage III single crystal data is available, have a body-centered cubic crystalline structure. Since the polycrystalline quasi-static data described in Chapter III demonstrated that both of these elements possessed stress-strain functions in agreement with the generalized behavior of

Eq. (3.1), it is of particular interest, because of the difference in their crystal structure, to examine their single crystal behavior.

If the stress and strain ratios of the TAYLOR (1938) and BISHOP and HILL (1951) aggregate theory are to apply to these body-centered cubic elements, one would expect that the value of $\overline{m}$ would be different because of the differences in the number of slip planes operable and the fact that the resolved shear deformation takes place on a different slip plane. In Figs. 4.27 and 4.28 are shown 4 low purity iron single crystal experiments of TAYLOR (1934) and 2 somewhat higher purity iron single crystal experiments of DOHI (1960). As may be seen from all 6 of the experiments the τ^2 vs γ plots are parabolic.

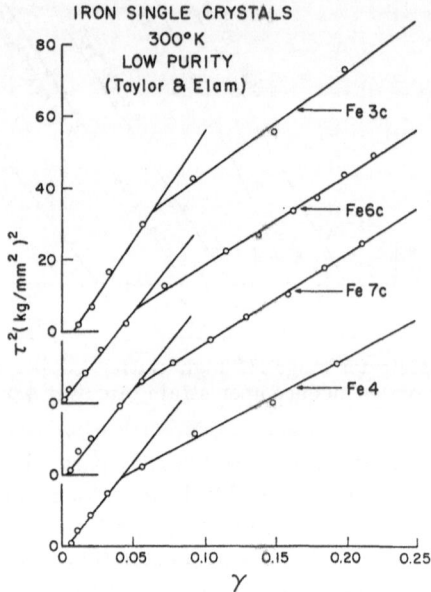

Fig. 4.27. τ^2 vs γ plots for low purity iron single crystals. Note the transition at the first single crystal transition strain of 4.6%

The $\beta_{ro}(II)$ and $\beta_{ro}(III)$ zero-point parabola coefficients for these experiments are tabulated among the single crystal data in Appendix I. Averaged values for each of the indicated mode indices r are compared in Table VIII with predicted values from Eq. (4.11), using a value of $\overline{m} = 3.06$. The correlation of these iron single crystal data with prediction is, of course, of considerable interest in itself. However, it is thought to be of particular significance that when the transitions occur in the data of TAYLOR (1938) shown in Fig. 4.27, the transition strain for each of the experiments is close to the value for the aluminum single crystal of $\gamma_N = 0.046$.

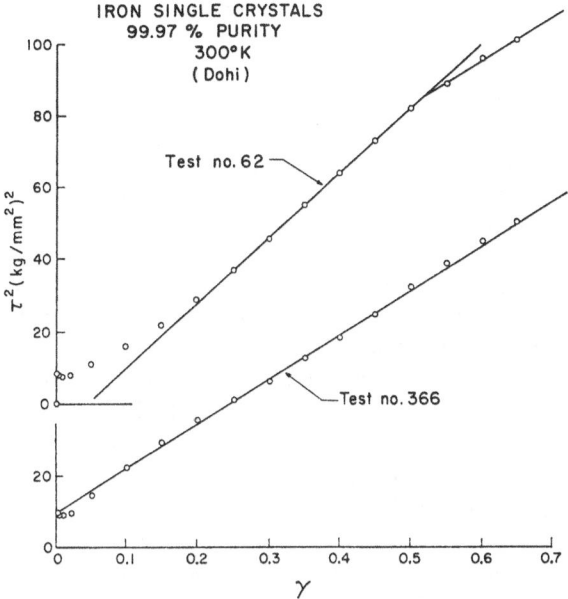

Fig. 4.28. τ^2 vs γ plots in higher purity iron single crystals. Note the transition at the fourth transition strain of 52.6%

Table VIII. *Averaged Fe data from Appendix I*

Mode Index γ	No. of Measurements	Average $\beta_{ro}(II)$ Experiments	Theoretical $\beta_{ro}(II)$	No. of Measurements	Average $\beta_{ro}(III)$ Experiments	Theoretical $\beta_{ro}(III)$
2	—	—	—	4	27.96	29.92
4	—	—	—	3	19.47	19.94
5	1	15.95	16.28	2	16.73	16.28
6	—	—	—	2	13.25	13.30

Thus, for body-centered cubic iron, not only are parabolas for the large distortional deformation predictable from the face-centered cubic single crystal data, i.e., $\overline{m} = 3.06$, but also the transition structure is independent of crystal structure.

Tantalum

The 26 single crystal experiments of MITCHELL and SPITZIG (1965) are in body-centered cubic tantalum. The range of ambient temperatures for which stage III deformation was observed is from 77° K to 573° K. Because the melting point of tantalum is high ($T_m = 3{,}269°$ K) the range of fractional melting point temperatures, T/T_m, is only from 0.0235 to 0.175. Most of the experiments were performed at a strain rate of $9.3 \times 10^{-4} \sec^{-1}$ although one series of experiments at 373° K were performed over a range of strain rates from $\dot{\varepsilon} = 9.3 \times 10^{-6} \sec^{-1}$ to $\dot{\varepsilon} = 9.5 \times 10^{-1} \sec^{-1}$. No dynamic data in tantalum polycrystals or single crystals is available. However, the quasi-static polycrystalline experiments of BECHTOLD, WESSEL, and FRANCE (1961) and of JAFFEE, MAYKUTH, and DOUGLASS (1961), shown in Figs. 3.12 and 3.13 in Chapter III furnished data in agreement with the polycrystalline portion of the parabolic generalization. MITCHELL and SPITZIG (1965) have provided tabulated values of θ_{II} and τ^* from which the $\beta_{ro}(II)$ calculations given in Appendix I could be determined. Fortunately they have also given the actual resolved shear stress, shear strain diagrams of the stage III data itself from which $\beta_{ro}(III)$ parabola coefficients could be directly determined. Averaged values for each mode index r are compared empirically with polycrystalline prediction in terms of Eq. (4.11) using $\overline{m} = 3.06$ in Table IX.

Table IX. *Averaged Ta data from Appendix I*

Mode Index r	No. of Measurements	Average $\beta_{ro}(II)$ Experiments	Theoretical $\beta_{ro}(II)$	No. of Measurements	Average $\beta_{ro}(III)$ Experiments	Theoretical $\beta_{ro}(III)$
2	—	—	—	2	25.715	24.44
4	1	16.86	16.29	3	16.56	16.29
5	—	—	—	1	14.28	13.30
6	15	10.99	10.86	8	10.62	10.86
7	3	9.05	8.87	13	8.88	8.87
8	2	7.94	7.24	6	7.41	7.24
9	1	6.00	5.91	5	5.85	5.91
10	—	—	—	3	5.11	4.82

The τ^2 vs γ plots of the data are shown in Figs. 4.29, 4.30, 4.31, and 4.32.

Despite differences in the initial aspects of the deformation for many of these tantalum single crystals when compared with the face-centered cubic crystals or with body-centered cubic iron, discussed above, the straight line portions of the large deformation not only provide evidence of parabolicity in the resolved shear stress, resolved shear strain functions, but also provide experimental zero-point parabola coefficients in agreement with prediction [Eq. (4.11)].

The fact that body-centered cubic tantalum and body-centered cubic iron are both quantitatively and qualitatively empirically determinable through the aggregate theory ratios in the same manner as

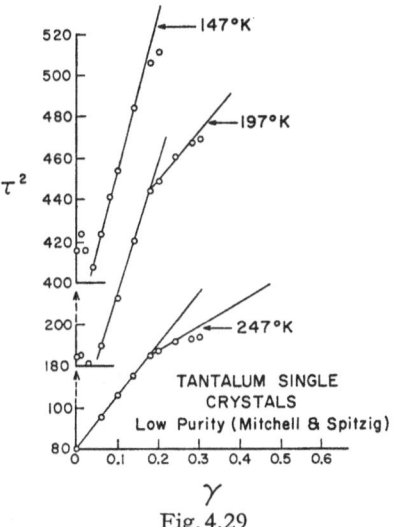

Fig. 4.29

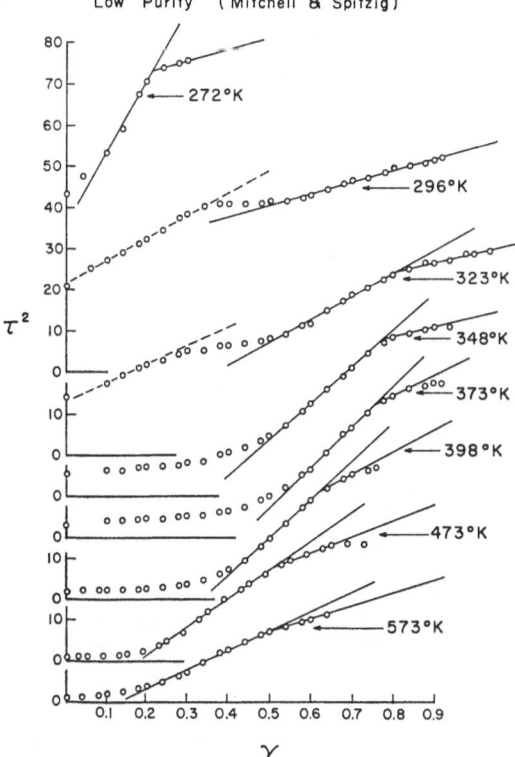

Fig. 4.30

face-centered cubic single crystals, i.e., $\overline{m} = 3.06$ for both crystal structures, emphasizes the need for reexamination of the aggregate theory with respect to crystal structures.

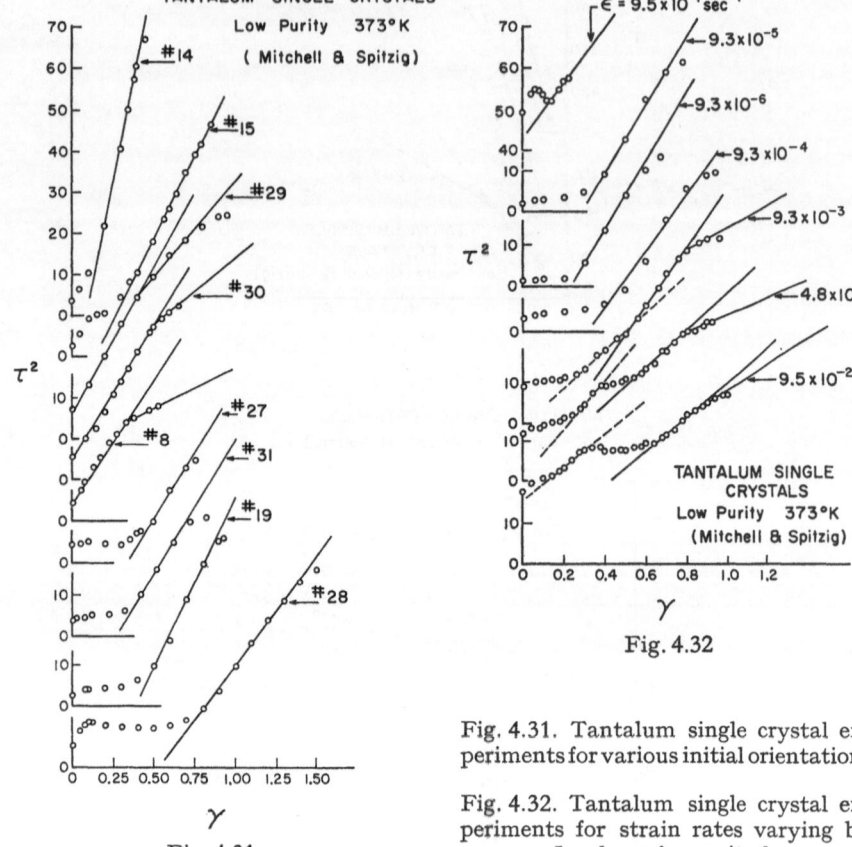

Fig. 4.31

Fig. 4.32

Fig. 4.31. Tantalum single crystal experiments for various initial orientations

Fig. 4.32. Tantalum single crystal experiments for strain rates varying by 5 orders of magnitude

Summary of Single Crystal Data

The total number of single crystal experiments in 8 crystallographically cubic elements is 430, of which 55 experiments in aluminum have been performed in the writer's laboratory, the remainder being from the literature between 1923 and 1967. The 55 experiments of the writer were individually compared with prediction by introducing the linearly temperature dependent parabolic generalized polycrystalline stress-strain function, Eq. (2.20), into the stress and strain ratios of the

TAYLOR (1938) aggregate theory, Eq. (4.8). The agreement of these experiments (see Figs. 4.6, 4.7, 4.8, 4.10, 4.13, 4.18, and 4.20) with the present generalization, is indeed remarkable, and emphasizes the fact that the origin of the aggregate behavior is derivable from a consideration of the behavior of the individual component.

When one includes the wave propagation single crystal data of Fig. 4.11, the range of strain rates in these aluminum single crystal measurements is from $\dot{\varepsilon} = 10^{-9} \sec^{-1}$ to $\dot{\varepsilon} = 10^3 \sec^{-1}$, a range of strain rates of 10^{12} for which the same resolved stress-strain function is applicable in the single crystal.

Despite the fact that specimen purities varied from 99.16% to 99.999%, that specimen diameters varied from 1.2 mm to 25.4 mm, that specimen length to diameter ratios varied from 3 to 89, that both tension and compression experiments were considered, that initial crystallographic orientations were widely variant, including what formerly have been considered to have been multiple slip orientations, and that two different crystal structures—face-centered cubic, and body-centered cubic (the latter from others' experiments)—are considered, there is quantitative and qualitative unity in the large distortional deformation of crystalline solids. The polycrystalline generalization of Eq. (3.1) and its single crystal stage III counterpart, Eq. (4.11), prescribe that finite distortional deformation occurs in the form of a series of discrete parabolic functions, designated by the mode index r. In terms of the proportionality of the parabola coefficient to the zero-point isotropic elastic shear moduli and the known fractional melting point, the large distortional deformation of all 19 elements and the several binary combinations may be considered as a single common solid.

Reduced to a common reference solid, the 375 single crystal experiments in 8 cubic elements tabulated in Appendix I are combined to demonstrate how closely the experimental $\beta_{ro}(\text{II})$ and $\beta_{ro}(\text{III})$ experimental zero-point parabola coefficients correlate with the predicted discrete mode index spectrum. Since transitions occur in much of these experimental data, the total number of zero-point parabola coefficients which have been calculated from these observations, 660, exceeds the total number of experiments, 375. Many of the experiments from the literature do not exhibit either a stage I or a stage II deformation and thus no $\beta_{ro}(\text{II})$ is determinable. In still other experiments no actual stage III data were provided so that there is no directly measured $\beta_{ro}(\text{III})$. Nonetheless, a comparison of 212 $\beta_{ro}(\text{II})$ zero-point parabola coefficients, and 269 $\beta_{ro}(\text{III})$ zero-point parabola coefficients experimentally establishes Eqs. (4.5) and (4.6) interrelating stage II and stage III deformation.

All the literature data of Appendix I are presented in terms of the mode index distribution in Table X and Table XI by comparing the zero-point reference solid parabola coefficients $\beta_{ro}(\text{II})/\mu(0)$ and $\beta_{ro}(\text{III})/\mu(0)$ with prediction. One notes from Eq. (4.11) for $T/T_m = 0$ that $\beta_{ro}/\mu(0) = (2/3)^{r/2} B_0/\overline{m}^{3/2}$. Table X includes all zero-point parabola coefficients, whether of the initial parabola or for subsequent transitions. In Table XI are compared zero-point parabola coefficients for initial parabolas only.

Table X. *Initial and transition parabolas*

Mode Index r	No. of Measurements	Theoretical $\beta_{ro}/\mu(0)_1$	Average Experiments $\beta_{ro}(\text{II})/\mu(0)$	Average Experiments $\beta_{ro}(\text{III})/\mu(0)$	Average All Experiments $\beta_{ro}/\mu(0)$
1	10	0.00427	0.00426	0.00411	0.00424
2	25	0.00349	0.00353	0.00337	0.00346
3	44	0.00285	0.00287	0.00278	0.00283
4	120	0.00232	0.00228	0.00227	0.00228
5	155	0.00190	0.00193	0.00190	0.00191
6	101	0.00155	0.00158	0.00155	0.00155
7	77	0.00127	0.00129	0.00129	0.00129
8	51	0.00103	0.00108	0.00103	0.00104
9	33	0.00084	0.00082	0.00086	0.00086
10	30	0.00069	0.00074	0.00071	0.00071
11	12	0.00056	—	0.00057	0.00057
12	2	0.00046	—	0.00045	0.00045

Table XI. *Initial parabolas only*

Mode Index r	No. of Measurements	Theoretical $\beta_{ro}/\mu(0)$	Average Experiments $\beta_{ro}(\text{II})/\mu(0)$	Average Experiments $\beta_{ro}(\text{III})/\mu(0)$	Average All Experiments $\beta_{ro}/\mu(0)$
1	10	0.00427	0.00426	0.00411	0.00424
2	25	0.00349	0.00353	0.00337	0.00346
3	43	0.00285	0.00287	0.00278	0.00283
4	112	0.00232	0.00228	0.00226	0.00227
5	152	0.00190	0.00193	0.00190	0.00191
6	72	0.00155	0.00158	0.00155	0.00156
7	41	0.00127	0.00129	0.00130	0.00130
8	17	0.00103	0.00108	0.00106	0.00107
9	8	0.00084	0.00082	0.00087	0.00085
10	3	0.00069	0.00074	0.00074	0.00074
11	2	0.00056	—	0.00057	0.00057

Both of these tables provide evidence that the stage III deformation of the crystal physics literature, and stage III prediction from stage II deformation, is in remarkable agreement with the present generalization. The fact that Table X includes transition parabolas as well as initial parabolas (both, of course, are tabulated in Appendix I) furnishes additional evidence of the quantitative predictability of the transition structure.

The writer's single crystal data and the data from the literature (the latter being included in Appendix I and Tables X and XI) have initial orientations which cover the entire stereographic triangle, including all three corners and lines of symmetry. In every instance, to obtain agreement it is necessary to compare data which has been resolved for macroscopic single slip. As was indicated earlier in the present chapter, it was as a consequence of this conclusion of the writer that BELL and GREEN (1967) undertook a study of the double slip deformation hypothesis which resulted in showing that the conclusion affirming the applicability of the present writer's macroscopic single slip parabolic generalization is, in fact, the actual behavior.

Of particular interest are the corner orientation experiments of KINGMAN, GREEN, and POND (1963), of LÜCKE and LANGE (1952), of STAUBWASSER (1954), and of KARNOP and SACHS (1927), for which unresolved axial data is available so that the data may be calculated in single slip. A number of experimentalists: PRICE and KELLY in copper — 1.8 wt. % beryllium (1963), PRICE and KELLY in aluminum alloy crystals (1964), and MURPHY and CALNAN in α-brass (1955) have measured crystallographic angles during deformation by means of x-ray diffraction for initial $\langle 100 \rangle$ orientations. In every instance in the $\langle 100 \rangle$ corner data rotation is towards the $\langle 101 \rangle$ position, as is expected for single slip.

An inspection of the initial orientation of the KINGMAN, GREEN, and POND (1963) compression data listed in Appendix I permits of the averaging of parabola coefficients for initial orientations in each of the three corners of the stereographic triangles. Combining these data with those of LÜCKE and LANGE (1952) and STAUBWASSER (1954), one obtains for orientations in the vicinity of the $\langle 100 \rangle$ corner $\beta_{ro} = 4.87$ kg/mm² for 25 measurements, which may be compared with the predicted value for $r = 6$ of $\beta_{ro} = 4.82$ kg/mm². For the $\langle 111 \rangle$ corner, the average of 15 experiments provides an experimental zero-point parabola coefficient of $\beta_{ro} = 5.69$ kg/mm² which may be compared with the predicted value of $\beta_{ro} = 5.90$ kg/mm².

The 13 experiments in the $\langle 110 \rangle$ corner of KINGMAN, GREEN, and POND (1963) fall into two groups: 8 experiments having an average experimental parabola coefficient of $\beta_{ro} = 6.88$ kg/mm², and 5 experiments providing an experimental parabola coefficient of $\beta_{ro} = 5.89$ kg/mm², which may be compared with predicted values of $\beta_{ro} = 7.24$ kg/mm² and $\beta_{ro} = 5.90$ kg/mm², for $r = 4$ and $r = 5$ respectively.

KARNOP and SACHS (1927) in experiment No. 29 have provided data within 3° of the $\langle 100 \rangle$ corner in which 8 slip systems have formerly been considered operable with no rotations assumed during deformation. Since the σ—ε axial data were provided, it is possible to calculate resolved data for this experiment in 99.1% purity aluminum at room temperature. This measurement is of particular interest in view of the fact that all of the other corner orientation data were of relatively high purity. As may be seen from Fig. 4.33, this experiment provided a resolved stress-strain function in close agreement with the present writer's prediction.

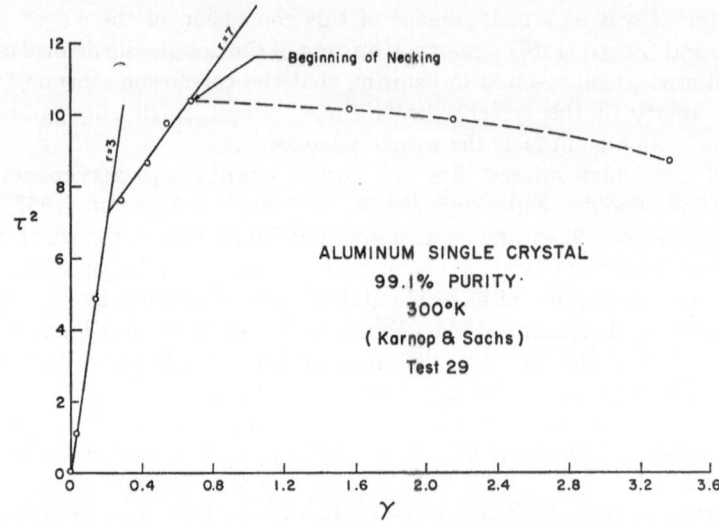

Fig. 4.33. τ^2 vs γ plot in a low purity aluminum single crystal whose initial orientation is within 3° of the ⟨100⟩ corner

Thus it has been shown that the resolved deformation of the single crystal when calculated as macroscopic single slip is independent of the initial orientation and independent of whether or not subsequent rotation passes a symmetry line. The curious fact in this experimental conclusion, however, is that the resolved stage III deformation of interest is well known from surface observations to be a multiple slip deformation. The fact that measured x-ray diffraction angles φ and λ for this multiple slip stage III deformation *are* determinable from the single slip, Eqs. (4.12) and (4.13), suggests that some sort of a complex averaging of the various slip systems is occurring which is independent of the initial orientation. Therefore, since one cannot be certain that the actual resolving procedures are understood, it is of interest to reconsider the single crystal deformation in terms of the axial stress and axial strain which are known from the experimental measurements. From the fact of the dependence of axial stress and strain values upon orientation, the stage III portion of the single crystal deformation may be written as a modification of Eq. (4.17). Independent of whether or not the kinematics of single slip actually is applicable, when it is used *all* initial orientations provide a common finite deformation function. From this unity one may empirically state a common functional relation for measured axial stress and strain in terms of initial orientation and independent of any kinematical consideration. One must of course, know from x-ray diffraction the angles φ_{ob} and λ_{ob} at the resolved shear strain γ_b of Fig. 4.1. In these terms Eq. (4.17) applies to all stage III deformation and we may write

$$\sigma_s = \beta_s \frac{(1 + \varepsilon_s - \varepsilon_b)}{(\text{Cos } \varphi_{ob})^{3/2}} \left\{ \frac{(1 + \varepsilon_s - \varepsilon_b)^2 + \text{Sin}^2 \lambda_{ob}}{[(1 + \varepsilon_s - \varepsilon_b)^2 - \text{Sin}^2 \lambda_{ob}]^{1/2} - \text{Cos } \lambda_{ob}} \right\}^{1/2}. \qquad (4.18)$$

If, as in low purity deformation or in high purity deformation with corner orientation, $\gamma_b = \varepsilon_b = 0$ and stage III deformation is zero at the origin, then the orientation angles of Eq. (4.18) become the initial angles of orientation; i.e., $\varphi_{ob} = \varphi_o$ and $\lambda_{ob} = \lambda_o$.

$$\sigma_s = \beta_s \frac{(1 + \varepsilon_s)}{(\text{Cos } \varphi_o)^{3/2}} \left\{ \frac{(1 + \varepsilon_s)^2 + \text{Sin}^2 \lambda_o}{[(1 + \varepsilon_s)^2 - \text{Sin}^2 \lambda_o]^{1/2} - \text{Cos } \lambda_o} \right\}^{1/2}. \qquad (4.19)$$

The subscript s has been introduced to distinguish between the uniaxial nominal stress and strain of the polycrystal and the uniaxial nominal stress and strain of the single crystal. From Eq. (4.11) the single crystal parabola coefficient is

$$\beta_s = (2/3)^{r/2} \mu(0) \frac{B_0}{m^{3/2}} (1 - T/T_m). \qquad (4.20)$$

Eq. (4.18) applies to all initial orientations and represents a measured stage III deformation, independent of resolving procedures for all 8 single crystal elements described in this chapter.

It can further be shown that since the polycrystalline parabola coefficients and single crystal parabola coefficients, when calculated for single slip, are related in terms of the aggregate theory, the ratio of axial stresses and axial strains of polycrystals and single crystals also may be stated independent of resolving procedures. These ratios are shown in Eqs. (4.21) and (4.22)

$$\frac{\sigma}{\sigma_s} = \overline{m} \text{ Cos } \varphi_o (\text{Cos } \lambda)^{1/2} \qquad (4.21)$$

$$\frac{\varepsilon_s}{\varepsilon} = \overline{m} \text{ Cos } \varphi_o \frac{\text{Cos } \varphi_o - \text{Cos } \varphi}{\text{Cos } \varphi_o \text{ Cos } \lambda - \text{Cos } \varphi \text{ Cos } \lambda_o}. \qquad (4.22)$$

It should be emphasized that Eqs. (4.18), (4.19), (4.21), and (4.22) represent experimental observations in terms of measured valus of σ, σ_s,

ε, ε_s, φ_0, λ_0, φ, and λ. Eq. (4.18) has the advantage of being independent of the resolution kinematics.

The experimental data of Appendix I and of Tables I through XI have been given as zero-point parabola coefficients by dividing measurements at each temperature by $(1 - T/T_m)$. The correlation of the zero-point data with prediction, therefore, experimentally establishes the linear temperature dependence for a fixed mode index r. The mode index of the initial parabola for these data does vary with temperature and purity. (High purity has been arbitrarily defined as 99.99% or higher

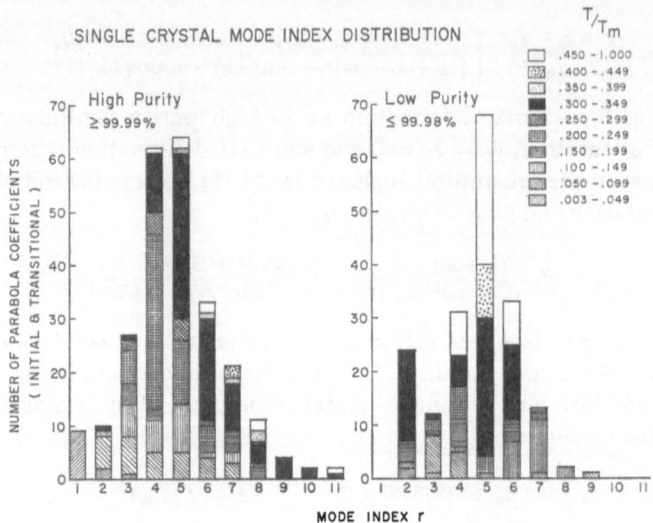

Fig. 4.34. The parabola coefficient mode index distribution as a function of fractional melting point temperature for all of the high and low purity single crystal data of Chapter IV

and low purity as below 99.99%.) Plots of mode index distributions for all of the tabulated initial single crystal zero-point parabola coefficients of Appendix I and Figs. 4.6 through 4.10, for fractional melting point intervals of 0.05 are shown in Fig. 4.34. The data are grouped for each purity so that the trend of initial mode indices increasing with temperature and with purity may be seen. As was shown in Chapter III, the initial polycrystalline zero-point parabola coefficients are similarly distributed with temperature and purity.

High purity aluminum polycrystals or multicrystals have relatively large grains when compared with dead annealed extruded commercial purity rod. To be certain that systematic orientations were not provided in forming these specimens from the melt, a 99.99% purity aluminum multicrystal was deformed in compression with x-ray diffraction determination of the orientation of individual grains before and after deformation. The specimen was initially etched and a sample grain was selected

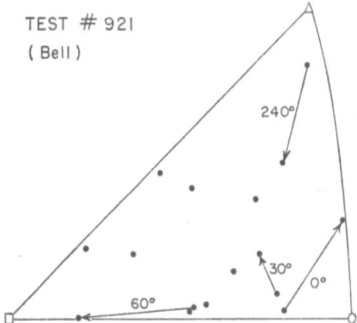

Fig. 4.35. Initial and final orientations determined by x-ray diffraction measurements at every 30° around an aluminum columnar multicrystal, test 921 of Fig. 4.36

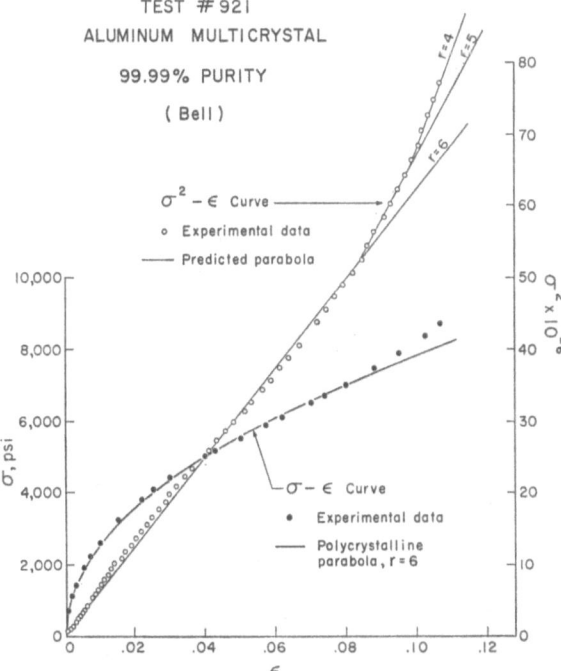

Fig. 4.36. $\sigma - \varepsilon$ and $\sigma^2 - \varepsilon$ plots of the high purity columnar multicrystal compression experiment whose initial and final orientations were shown in Fig. 4.35

near the center along every 30° genetrix. The initial orientations shown in Fig. 4.35 demonstrate that even though the specimen of experiment 921 was tested as poured in a cylindrical mold, the crystallographic orientations were randomly distributed. Also shown in Fig. 4.35 are final orientations, where determinable, which include both crystallographic and rigid body rotations of the individual grains. They demonstrate that at large deformation at a final critical transition strain of 11.5% the combined rotation has no set direction in compression common to all grains. The σ—ε and σ^2—ε plots of this experiment are shown in Fig. 4.36 from which the usual $r = 6$ parabola of this high purity multicrystal at room temperature is seen to govern the finite deformation.

Chapter V

The Discrete Distribution of Zero-Point Isotropic Elastic Shear Moduli of the Elements

The finite distortional deformation modes and their transition properties are characterized by the integer r. What is of interest in the present chapter is the closely related experimental fact that the zero-point isotropic elastic shear moduli $\mu(0)$ are also distributed in a similar discrete pattern among the elements.

The experimental parabola coefficients of different crystalline solids which are proportional to the zero-point linear elastic isotropic shear modulus $\mu(0)$ occur in numerical ratios of precisely the same form and magnitude as the deformation modes characterizing the transition behavior of any one crystalline solid. This empirical relation is given in Eq. (5.1), where $s = 1, 2, 3, ----$ and $p = 0$ or 1.

$$\mu(0) = (2/3)^{s/2} (2/3)^{p/4} A \tag{5.1}$$

where $A = 2.89 \times 10^4 \, kg/mm^2$.

The calculated distributions of Eq. (5.1) are compared in Table A, Appendix II, with averaged experimental values of $\mu(0)$ for 48 elements and two binary combinations.

Although the observed distribution represents a correlation with measured large deformation parabola coefficients, it is interesting to test the agreement between the predicted and experimental distributions independent of that correlation. The ratio of χ^2 summation for 96 values evenly distributed from the plus to minus half intervals between predicted values to that for the 96 measurements of Table A, Appendix II, (Eq. 5.2) is 2.25.

$$\chi^2 = \sum_{i=1}^{96} \left(\frac{\mu_i(0) \, \exp - \mu_i(0) \, \text{pred}}{\mu_i(0) \, \text{pred}} \right)^2. \tag{5.2}$$

Although this is a modified χ^2 test, the ratio is sufficiently large to establish the distribution independent of the large deformation correlation from which it was originally predicted.

An examination of the average distribution over the entire range reveals that of the total of 50 crystalline solids, 28 occur in 8 of the

44 predicted values between $s = 1$, $p = 0$ and $s = 22$, $p = 0$; i.e., between the highest observed values of Os and Ir and the lowest for Cs. None of the eight groups has less than 3 members. For the averaged members of these eight groups, the percentage deviation from prediction is 0.44%, the largest deviation from prediction being 1.96% for the group containing bismuth. Ten elements have 3 or more separate measurements, providing an average for which a closer correlation can be expected. Of these 10 elements, the average percentage deviation from prediction is 1.1%, with none exceeding 2.1%. The plus one-half interval is $+5.4\%$ of prediction and the minus one-half interval -4.9% of prediction. Thus these multiple data for both the grouping of the elements and the individual elements with more than three measurements provide a good correlation with the zero-point isotropic shear moduli distribution of Eq. (5.1).

In addition to the directly measured shear moduli of Table A, Appendix II, indirectly determined values may be obtained from measured isotropic Young's moduli and bulk moduli, E and K, in terms of Eq. (5.3).

$$\mu = \frac{3KE}{9K-E}. \tag{5.3}$$

Directly measured $\mu(0)$ data of Table A, Appendix II have been compared with indirectly measured $\mu(0)$ by means of Eq. (5.3). Despite the fact that ratios of large numbers are involved, the agreement where comparisons could be made was sufficient to warrant the inclusion of such indirect data for 10 elements. The indirect calculations for 10 elements for which no directly measured $\mu(0)$ are available, are shown in Table B, Appendix II.

In Fig. 5.1 is shown the spectral distribution of the 58 elements and two binary combinations of Tables A and B, Appendix II, compared with the predicted distribution Eq. (5.1). The data are shown in a log plot to emphasize agreement throughout the large range of values. For greater clarity, both experimental and predicted $\mu(0)$'s are multiplied by $(2/3)^{p/4}$ in Fig. 5.1, thus combining the two distributions into one.

In general, isotropic shear moduli are available in the literature only at room temperature. In order to obtain zero-point isotropic shear moduli for as many of the elements as possible, a study was made of the temperature distribution between $4°$ K and $300°$ K; i.e., room temperature, for elements for which such data existed. Upon establishing that a uniform empirical behavior existed for this group, the same distribution was assumed for the remainder of the elements. For 23 of the elements the melting point is sufficiently high so that the fractional melting point temperature T/T_m is small when $T = 300°$ K, with the result that the room temperature value of μ nearly coincided with $\mu(0)$, thus providing an additional check of the empirical distribution of Eq. (5.1).

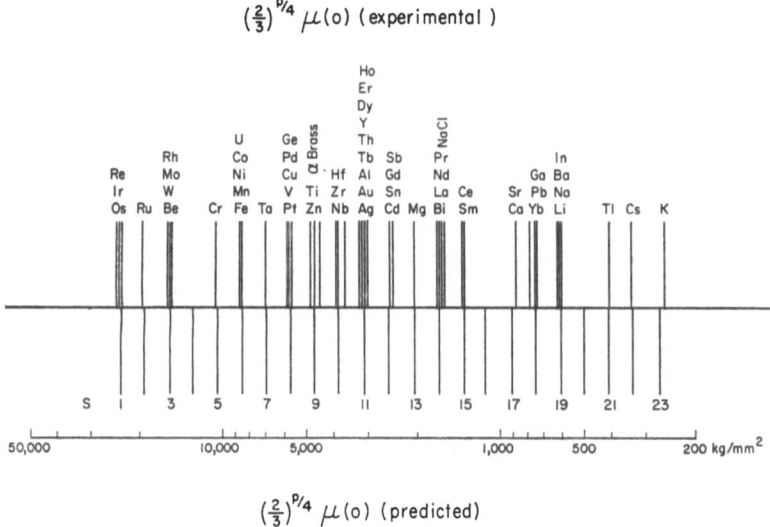

Fig. 5.1. A comparison of averaged experimental $\mu(0)$ with distribution of Eq. (5.1). Separate distributions for $p = 0$ and for $p = 1$ are shown as a single distribution by comparing the predicted $(2/3)^{p/4} \mu(0)$ with experiment

The Experimental Data

As of the present writing, the generalized large deformation behavior of 27 crystalline solids has been studied by the writer and his students (19 elements and 8 binary combinations). Available in the literature are the temperature distributions of shear moduli between 4.2° K and room temperature of 7 members of this group; i.e., Al, Cu, Au, Ag, Pb, Mg, and Ni. It was from the comparison of the dynamic large deformation parabola coefficients of these elements that the discrete distribution of zero-point isotropic shear moduli was first observed. In particular, aluminum, gold, and silver have identical experimental parabola coefficients although melting points, mass densities, etc., are widely different.

In a search for a common parameter relating the large deformation of these three elements, other than the fact that all three were face-centered cubic, two were found: the lattice spacing, and the zero-point isotropic shear modulus. Nickel and iron also were observed to have identical large deformation parabola coefficients and, in addition, have nearly identical lattice spacing and zero-point isotropic shear moduli; hence, it at first seemed that in some manner both of these parameters would have to be considered simultaneously. However, copper, whose parabola coefficients are far below those of nickel and iron, while possessing nearly the same lattice spacing as these two, has likewise a much lower zero-point isotropic shear modulus. The ratio of parabola coefficients and the ratio of zero-point isotropic shear moduli were experimentally the same. The ratio of the zero-point isotropic shear modulus of copper to that of aluminum, silver, and gold was also found to be the same as the ratio of the corresponding large deformation parabola coefficients, as was the situation when nickel and iron data were compared with aluminum, silver, and gold (BELL, 1965a). These observations led to the study of the zero-point isotropic shear moduli, not only for the remaining members of the group for which finite deformation data was available, but also for all of the remaining elements for which isotropic shear moduli were known.

Most elastic isotropic shear modulus determinations have been made at room temperature, whether directly measured or calculated from the bulk moduli and Young's moduli measurements; hence, some method had to be found to extrapolate such data to the vicinity of absolute zero.

All but one of the shear moduli measurements as a function of temperature found in the literature have been made on single crystals. A study of the experimental data of the several investigators who have provided measurements of the C_{44} shear modulus vs temperature for the 7 elements (Al, Cu, Ag, Au, Pb, Mg, and Ni) revealed that these data in the region between $4.2°$ K and room temperature fitted very closely to the empirical relations of Eq. (5.4).

$$\mu(T/T_m) = \mu(0) \qquad 0 \leq T/T_m \leq 0.06$$

$$\mu(T/T_m) = 1.03\,\mu(0)\,(1 - T/2T_m) \quad 0.06 \leq T/T_m \leq \frac{300}{T_m}. \tag{5.4}$$

Calculated values of $\mu(0)$ in terms of Eq. (5.4) based upon the experimental $\mu(T)$ data of the designated investigators are shown diagrammatically in Fig. 5.2. All data of Fig. 5.1 and of Table A and Table B of Appendix II, either are directly measured near $0°$ K or are extrapolated by Eq. (5.4).

Aluminum and magnesium are nearly isotropic so that $C_{44} \approx \mu$. The data of ZUCKER (1955) in Fig. 5.2 (triangles) were obtained for the polycrystalline metal. Twenty-three of the 53 elements with high melting point T_m for which direct measurements of μ are available have a value of $T/2T_m$ for T at room temperature of less than 0.09, which corresponds to room temperature shear moduli from 0.1% to 6.7% of the zero-point value.

The room temperature and zero-point isotropic shear moduli of these 23 elements are listed in Table C, Appendix II. They provide a close approximation to the discrete distribution of Fig. 5.1, Table A, Table B of Appendix II, and Eq. (5.1), independent of the temperature extrapolation of Eq. (5.4).

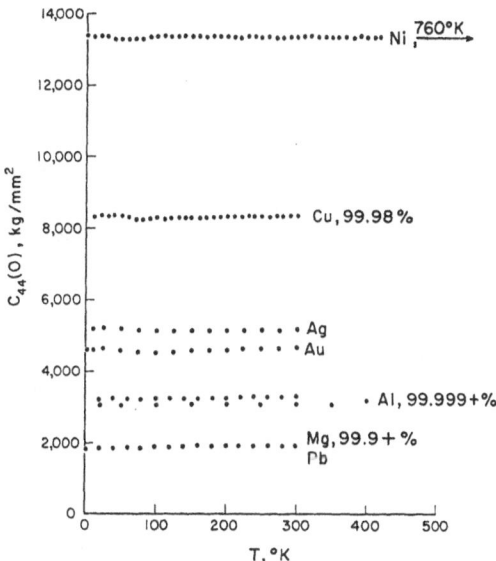

Fig. 5.2. Single crystal shear moduli $C_{44}(0)$, calculated from experimental data at various temperatures by Eq. (5.4). *Al* KAMM, G. N., and G. A. ALERS: J. Appl. Physics 35, No. 2, 327 (1964). — VALLIN, J., M. MONGY, K. SALAMA, and O. BECKMAN: J. Appl. Phys. 35, No. 6, 1825 (1964); Al (polycrystal). — ZUCKER, C.: J. Acoust. Soc. Amer. 27, No. 2, 318 (1955). *Au* GOENS, E. VON: Ann. Physik. 38, 456 (1940). — NEIGHBOURS, J. R., and G. A. ALERS: Phys. Rev. 111, Ser. 2, 707 (1958). *Ni* ALERS, G. A., J. R. NEIGHBOURS, and H. SATO: J. Phys. Chem. Solids 13, 40 (1960). *Mg* SLUTSKY, L. J., and C. W. GARLAND: Phys. Rev. 107, No. 4, 972 (1957). *Ag* NEIGHBOURS, J. R., and G. A. ALERS: Phys. Rev. 111, Ser. 2, 707 (1958). *Cu* GOENS, E. VON: Ann. Physik 38, 456 (1950). — OVERTON, W. C., Jr., and JOHN GAFFNEY: Phys. Rev. 98, No. 4, 969 (1955). *Pb* GOENS, E. VON: Ann. Physik 38, 456 (1940). — PRASAD, S. C., and W. A. WOOSTER: Acta Crystallogr. 9, 38 (1956)

If the 23 elements for which the room temperature value approximates the zero-point value, are added to the 7 elements for which direct measurements of the zero-point values and temperature distributions are available, then 29 of the 58 elements for which values of μ exist may be considered to have zero-point values established independent of the empirical temperature extrapolation of Eq. (5.4).

On comparing Eqs. (2.20), (3.1) and (4.11) with Eq. (5.1) it appears obvious that the integral indices r and s are different representations of the same phenomenon. The integral index r designates the large deformation mode. The value of r for the initial mode depends upon the ambient temperature and purity. Transitions from one large deformation mode to another occur at specified shear angles as the magnitude of the deformation increases. Because the deformation mode transitions, occurring at predictable shear angles, always change by multiples of $(2/3)^{1/2}$ independent of whether $p = 0$ or 1 for the shear modulus in Eq. (5.1), the index of Eq. (5.1) was chosen as $s/2 + p/4$ rather than as the single index $K/4$ where $K = 2s + p$.

It is interesting that the observed distortional transitions (BELL, 1965a; BELL and SUCKLING, 1962) are similar in form to the earlier hydrostatic transitions studied by BRIDGMAN (1931). For example, a hydrostatic transition of BRIDGMAN in nickel, shown in Fig. 5.3, is characterized by the same abrupt change in slope as the writer's τ^2 vs γ distortional data.

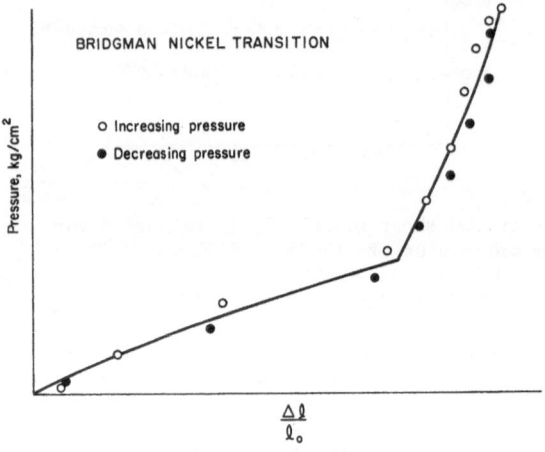

Fig. 5.3. A P. W. BRIDGMAN hydrostatic transition of the "second kind" in nickel

But, the present writer's distortional deformation transitions, which occur at large strain for relatively small stress, unlike the hydrostatic transitions which occur for large stress at relatively small strain, are not reversible. A detailed study of experimental isotropic elastic bulk moduli data revealed no such distribution among the elements as that shown here for the isotropic elastic shear modulus. Nevertheless, the fact that the decomposition of large distortional deformation into parabolic components is given by discrete changes in the zero-point isotropic shear moduli usually associated with infinitesimal deformation, suggests a common origin for both hydrostatic and distortional data transitions.

Experimentally measured isotropic linear elastic moduli for deformation at large prestress have been studied by the writer through the Portevin - le Chatelier effect. These data in polycrystalline aluminum which are described in a later chapter suggest that the index r and the index s are separate manifestations of the same phenomenon. A shift in measured moduli corresponding to a change in s has been observed at transition points at which a known change in r has taken place. In these terms r designates the discrete distribution of the moduli which occurs in any given element, and s that which is observed when comparing the initial very small deformation of the different elements.

The fact that the zero-point linear elastic isotropic shear moduli of 58 elements are arranged in the discrete distribution spectrum of Eq. (5.1) and are proportional to the large deformation initial and transition parabola coefficients of all of the crystalline elements and alloys thus far studied by the writer, indicates a high degree of order for finite deformation beyond the region in the immediate vicinity of the elastic limit, or critically resolved shear stress. The same moduli, of course, govern the linear elastic infinitesimal deformation below the elastic limit, or critically resolved shear stress. One may conclude that both nonlinear finite distortional deformation and infinitesimal linear elastic shear deformation have their origins in the same zero-point behavior. That the correlation between isotropic shear moduli and parabola coefficients is a zero-point phenomenon is further demonstrated by the differences in their temperature dependence. That the isotropic modulus is the significant parameter for both the anisotropic single crystal and the isotropic polycrystal should be particularly noted in assessing the non-dependence of finite distortional deformation upon specific crystallinity which is implied in the present generalization. It is interesting to examine these zero-point elastic isotropic shear moduli when plotted against atomic numbers, Fig. 5.4.

Because of the limited influence of differences in Poisson's ratio, such a plot is not too different from earlier similar plots of compressibility and atomic number (RICHARDS, 1915). One may examine earlier suggestions with respect to the calculation of elastic shear moduli from atomistic considerations. Earlier calculations (FUCHS, 1936; MOTT and JONES, 1936) have indicated that the shear modulus of the cubic elements should be proportional to the reciprocal space filling atomic volume Ω. A plot of the data of Fig. 5.1 and Tables A and B of Appendix II for $\mu(0)$ vs $N = 1/\Omega$ is shown in Fig. 5.5, as is a plot of $\mu(0)$ vs T_m in Fig. 5.6, and $\mu(0)$ vs $1/r_c^2$ in Fig. 5.7, where r_c is the closest atomic spacing.

10*

None of the comparisons in Figs. 5.5, 5.6, and 5.7 show other than a trend of higher moduli with increasing N, T_m, and $1/r_c^2$. In view of the high degree of numerical accuracy obtained when parabola coefficients

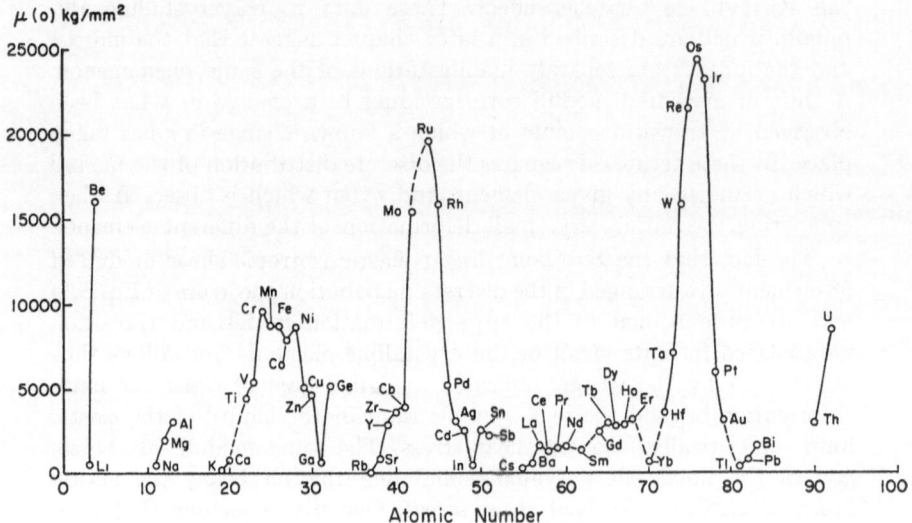

Fig. 5.4. The zero-point isotropic elastic shear moduli of the elements plotted against the atomic number

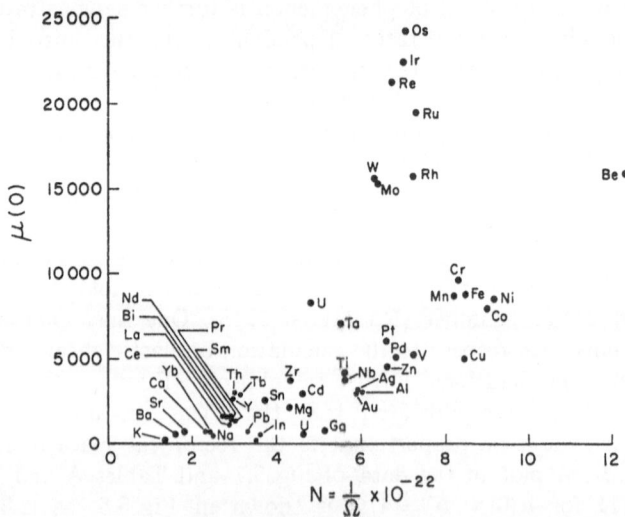

Fig. 5.5. The zero-point isotropic elastic shear moduli of the elements plotted against the reciprocal space filling atomic volume Ω

are compared, as the writer has described in Chapters II, III, and IV of the present monograph, the scatter in the plots of Figs. 5.5, 5.6, and 5.7 do not offer an encouraging point of departure for a theoretical treatment of the problem. Nevertheless, the discrete distribution of the

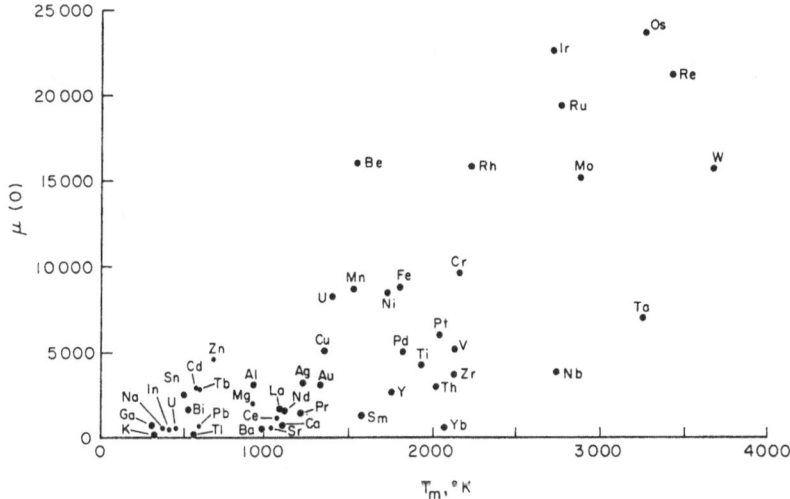

Fig. 5.6. The zero-point isotropic elastic shear moduli of the elements plotted against the melting point, T_m

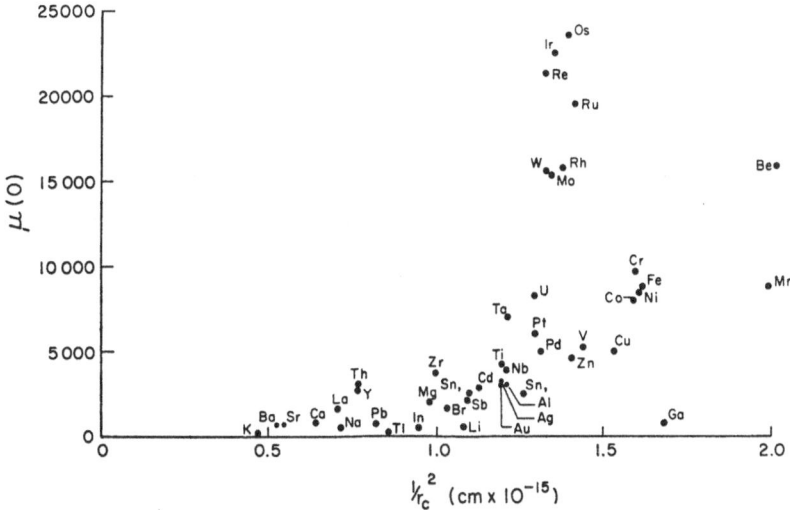

Fig. 5.7. The zero-point isotropic elastic shear moduli of the elements plotted against the reciprocal square of the closest atomic spacing, r_c

zero-point isotropic shear moduli for infinitesimal elastic deformation and the similarly distributed parabola coefficients for large distortional deformation should provide the experimental basis for a unified theory of distortional deformation not directly dependent upon the crystalline structural details of the individual element.

A satisfactory explanation of the finite distortional deformation mode and transition structure can be made only if one takes cognizance of the fact that the zero-point isotropic linear elastic shear moduli of 58 elements occur in the form of a discrete spectrum.

Chapter VI

Multiple Isotropic Linear Elastic Moduli

From the experimental research of COULOMB in 1784 who measured isotropic shear moduli in torsion, and the extensional experiments of YOUNG in 1807 who determined an E modulus in iron, to the present, a vast experimental literature has been written which has provided numerical values of elastic constants in nearly every crystalline solid. Measurements of isotropic polycrystalline moduli or of anisotropic single crystal constants in specified directions are reproducible from one experimentalist to another. The widely quoted measurements of E moduli and shear moduli of G. WERTHEIM obtained 125 years ago are numerically not too different from newly measured experimental values appearing in current journals and handbooks, despite the differences in experimental techniques and methods of specimen preparation.

To propose the existence of multiple values for isotropic linear elastic moduli in a single element, therefore, would certainly appear contrary to historical fact. Referring to the established values as stable moduli which may be determined for any crystalline solid at near zero stress, the multiple isotropic linear elasticity of interest in this chapter arises when the polycrystalline solid has been subjected to a prior dead annealed thermal history, or to a series of reversed loads at small strain which may or may not include small controlled amounts of plastic deformation, or when the polycrystalline solid is undergoing a finite distortional deformation with large plastic deformation. Under any of these circumstances a discrete predictable distribution of isotropic linear elastic moduli is observed. Indeed, numerous experimentalists from Lord KELVIN in 1880 to the present have commented upon the lowering of elastic moduli for fully annealed solids.

It has been shown above that, when the mode index r of the polycrystal or single crystal of one element is the same as that of another element, then for a given fractional melting point temperature T/T_m, parabola coefficients are in the ratio of the zero-point isotropic linear elastic shear moduli. This ratio (*see* Chapter V) is given by the respective integers, s and p, in the moduli distribution factor $(2/3)^{s/2} \times (2/3)^{p/4}$. The finite distortional deformation of a given element, as was shown in Chapters III and IV, undergoes a series of transitions with increasing deformation, which occur in precisely the same form for different integer values of the mode index r; i.e., $(2/3)^{r/2}$.

This striking parallel in the form of finite distortional deformation *among* the elements and among the finite deformation modes of an *individual* element suggested the desirability of a further study of these phenomena to determine whether or not the integers, r and s, were separate statements of the same phenomenon. This would mean that the quantity $(2/3)^{r/2}\mu(0)$ which occurs in both Eq. (3.1) for the polycrystal and Eq. (4.11) for the single crystal, could be replaced by the quantity from Eq. (5.1) of: $(2/3)^{s/2}(2/3)^{p/4}A$, where p would be unchanged for a given solid, and s would now not only include the stable value described in Chapter V but also would take on the multiple values of the mode index r.

The physical implication of this suggestion that an individual element may have multiple elastic moduli in a discrete distribution is that during finite deformation the small elastic distortion of the lattice produces second-order phase changes in the solid which result in a change of the elastic properties. The finite deformation parabolas, when such changes take place, would remain proportional to the applicable zero-point isotropic shear modulus in effect within the region of deformation of interest.

To establish whether or not multiple elastic regions or a discrete distribution of isotropic linear elastic moduli in a single element did exist, and if observed, to ascertain that they did fall in the predicted discrete distribution, a series of experiments were performed by the writer. As will be shown in the remainder of the present chapter, an overwhelming amount of experimental evidence has established that such multiple isotropic moduli do exist in a single element, paralleling the small stress stable isotropic elastic moduli distribution among the elements.

In 1955 the present writer first realized that for dead annealed metals the numerical value of isotropic elastic moduli obtained in dead-weight uniaxial stress experiments could vary widely from one experiment to another. At that time the first diffraction grating measurement of finite amplitude wave fronts had been obtained in completely annealed aluminum (BELL, 1956b). The problem was to compare such data with wave speed calculations obtained from the slopes of quasi-static uniaxial compression experiments in the same solid. The expected E modulus for aluminum at room temperature, determined from a Poisson's ratio of $1/3$ is 10.2×10^6 psi. This corresponds to a zero-point isotropic shear modulus for which $s = 11$ and $p = 0$ in Eq. (5.1). Observed moduli in dead-weight uniaxial compression experiments varied for completely annealed aluminum from 5.6×10^6 psi to 12.4×10^6 psi or for what is now known to be $s = 14$ to $s = 10$. The important point in these observations is not that one may modify metallurgical variables to produce moduli changes but that the moduli values obtained when this is done follow the same discrete distribution, Eq. (5.1), as pertains among the stable moduli of the various elements.

The writer's concern in 1955 over the variability of the elastic portion of dead annealed quasi-static measurements, suggested at the time the desirability of an independent check of this phenomenon. In 1956 the

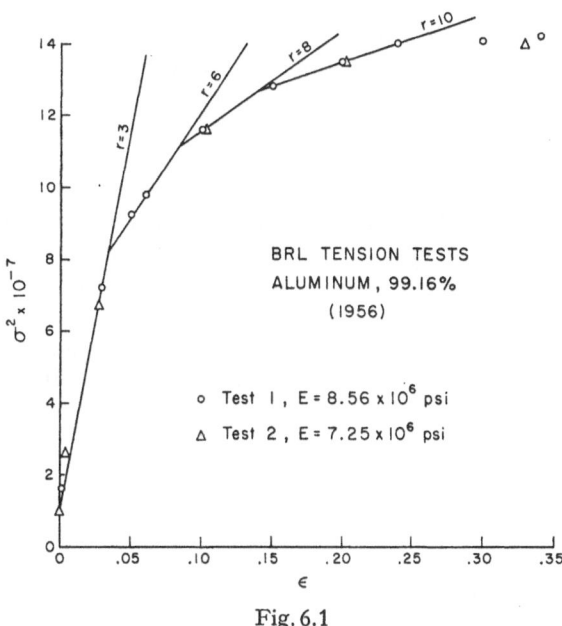

Fig. 6.1

U.S. Army Ballistics Research Laboratories, Aberdeen Proving Ground, carried out two tension experiments and two compression experiments on dead annealed aluminum polycrystals meeting the present writer's specifications. The large deformation portion of these data may be seen in the σ^2 vs ε plots of Fig. 3.4 as the BRL compression data, and in Fig. 6.1 for the two tension experiments. All four experiments provide predicted parabolas and finite deformation mode transitions. The E moduli for tension were $E = 8.56 \times 10^6$ psi and $E = 7.25 \times 10^6$ psi, instead of the stable value for $s = 11$, $p = 0$, of $E = 10.2 \times 10^6$ psi. One may compare the measured values with 8.35×10^6 psi for $s = 12$ and 6.8×10^6 psi for $s = 13$. One of the measured compression E moduli had the stable value for $s = 11$ of $E = 10.2 \times 10^6$ psi.

Lord KELVIN (1880) in describing an 1865 investigation of the effect on the torsional rigidity of wires of different metals produced by stressing them longitudinally beyond the limits of elasticity, commented on the highest and lowest rigidities found in copper as follows:

"Highest rigidity 473×10^6 g/cm², being that of a wire which had been softened by heating it to redness and plunging it into water, and which was found to be of density 8.91 ... Lowest rigidity 393.4×10^6 g/cm², being that of a wire which had been rendered so brittle by heating it to redness surrounded by powdered charcoal in a crucible and letting it cool very slowly that it could scarcely be touched without breaking it."

In copper, a room temperature modulus of 4,730 kg/mm² corresponds in terms of the temperature relations of Eq. (5.4), to a zero-point isotropic shear modulus $\mu(0) = 5,160$ kg/mm², which is precisely the predicted value of the stable zero-point isotropic shear modulus for copper for $s = 8$, $p = 1$ (see Chapter V). The predicted value of the room temperature shear modulus for $s = 9$, $p = 1$, corresponding to a unit change in s, is 3,860 kg/mm². This value is very close to the completely annealed modulus of 3,934 kg/mm² obtained by KELVIN over a hundred years ago in his torsion pendulum experiments.

Lord KELVIN (1880) also cites earlier experimental tensile E moduli measurements of WERTHEIM where, for hard-drawn copper, a value of 12,450 kg/mm² was obtained, and for annealed copper, 10,520 kg/mm². Both of these measurements were for quasi-static experiments. Their ratio is 1.19, or not quite $(3/2)^{1/2}$, similar to KELVIN's torsion measurements. The predicted value from Eq. (5.1) for $s = 8$, $p = 1$ is 12,600 kg/mm² and for $s = 9$, $p = 1$, 10,350 kg/mm². WERTHEIM also reports values of 12,510 kg/mm² and 12,540 kg/mm² for hard-drawn copper from experiments of transverse vibration and longitudinal vibration, respectively. For the annealed solid these values are 11,830 kg/mm² and 12,540 kg/mm² for the respective dynamic measurements. This observation of no change in dynamically measured moduli between hardened and annealed elements for which quasi-static moduli exhibit differences, is consistent with the present writer's observation in annealed aluminum. In aluminum, dynamic moduli determined from longitudinal wave speeds always provide the stable value.

In 1936 PHILLIPS and SMITH published a comparison of repeated elastic tension experiments in hard-drawn copper and in annealed copper. For a Poisson's ratio of $v = 0.34$ the stable zero-point isotropic shear modulus in copper, $s = 8$, $p = 1$ of Eq. (5.1), provides a room temperature E modulus of 17.95×10^6 psi, which is in very good agreement with the reported moduli of PHILLIPS and SMITH (1936) for hard-drawn copper wire, as may be seen in Fig. 6.2. The modulus for the first loading has a somewhat lower value of 17×10^6 psi. At a strain of 0.2% a change occurs to a modulus of 12.2×10^6 psi, which is close to the value of 11.95×10^6 psi for $s = 10$, $p = 1$. Subsequent loading and unloading provided the stable 17.9×10^6 modulus.

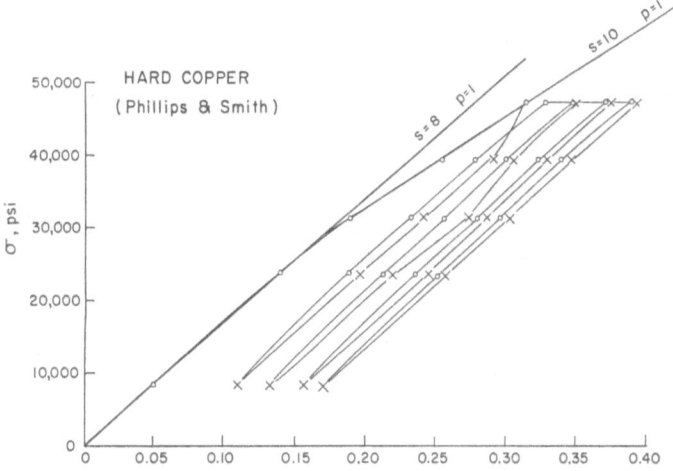

Fig. 6.2. Repeated loading of hard-drawn polycrystalline copper at small strain

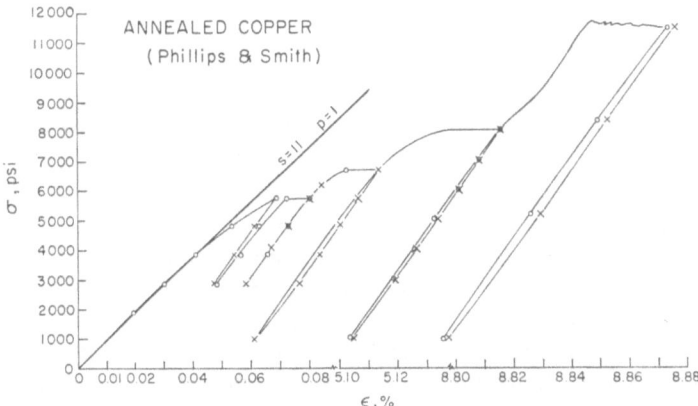

Fig. 6.3. Repeated loading of annealed polycrystalline copper at small strain; two loading and reloading cycles at large strain also are shown

The tension experiments of PHILLIPS and SMITH in annealed copper (Fig. 6.3) provided an initial modulus of 9.5×10^6 psi, closely corresponding to the value of $s = 11$, $p = 1$ of 9.7×10^6 psi instead of the stable modulus of hard-drawn copper of 17.95×10^6 psi for $s = 8$, $p = 1$. On the second and subsequent loading, unloading, and reloading at small strain, and following plastic strains as high as 5% and 9%, moduli were obtained which varied from 13.3×10^6 psi to 13.9×10^6 psi. These measurements involved lapsed times of many hours between one loading cycle and another. The average of these 9 measured moduli, 13.5×10^6 psi, for the successive loading and unloading of annealed copper lies between the values of $s = 9$ and $s = 10$, which is far below the 17.95×10^6 psi for the stable elastic modulus of $s = 8$, $p = 1$.

In 1937 McKeown and Hudson also commented upon the change in the elastic E modulus when comparing annealed and work-hardened stress-strain functions for Ag, Au, and Cu. These data are shown in Figs. 6.4, 6.5, and 6.6. These are the same experiments whose large deformation characteristics were described in Chapter III and shown in Figs. 3.14, 3.15, and 3.17. The work-hardened specimens were prestressed to 5% before the experiments were performed.

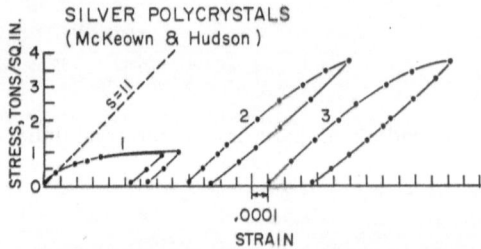

Fig. 6.4. *Curve 1* annealed (as received); curves for 0.1 and 0.01 ton/in.^{2}/min rates of stressing. *Curve 2* annealed (as received) $+ 5\%$ tensile overstrain. *Curve 3* annealed (as received) $+ 5\%$ tensile overstrain $+ \frac{1}{2}$ hr. at 350° C

In 1952 Richards presented a comparative study of isotropic elastic moduli in a single crystalline solid. The crystalline solid was 1.85% beryllium copper. The specimens were divided into three lots identified as H (0.091 in. diameter), J (0.219 in. diameter), and K (0.560 in. diameter). Each lot was divided into two groups in one of which the specimens were cold-drawn (designated as $\frac{1}{2}$ H) and in the second, precipitation-hardened (designated as $\frac{1}{2}$ HT). Similar sets of specimens were sent to a large number of laboratories which were asked to determine tension E moduli, compression E moduli, μ shear moduli, and Poisson's ratio. The E moduli and the shear moduli μ were found to have different values for the cold-drawn and the precipitation-hardened specimens, independent of the specimen diameter and independent of the type of test including whether it was static or dynamic.

The 134 E modulus measurements and the 41 μ modulus measurements using 16 different experimental methods including quasi-static,

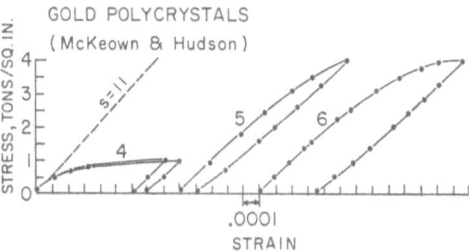

Fig. 6.5. *Curve 4* annealed (as received); curves for 0.1 and 0.01 ton/in.²/min rates of stressing. *Curve 5* annealed (as received) + 5⁰/₀ tensile overstrain. *Curve 6* annealed (as received) + 5⁰/₀ tensile overstrain + ¹/₂ hr. at 300° C

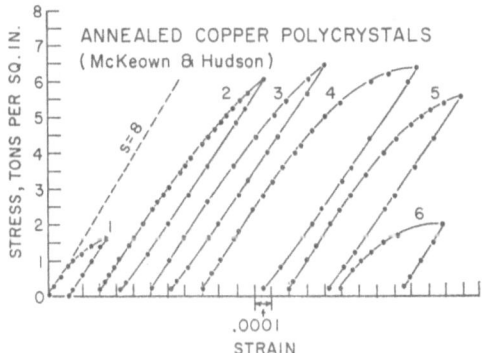

Fig. 6.6. *Curve 1* Hard-rolled and annealed 1 hr. at 500° C. *Curve 2* Hard-rolled, annealed 1 hr. at 500° C + 5⁰/₀ tensile overstrain. *Curve 3* Hard-rolled, annealed 1 hr. at 500° C + 5⁰/₀ tensile overstrain + 1 hr. at 200° C. *Curve 4* Hard-rolled, annealed 1 hr. at 500° C + 5⁰/₀ tensile overstrain + 1 hr. at 400° C. *Curve 5* Hard-rolled, annealed 1 hr. at 500° C + 5⁰/₀ tensile overstrain + 1 hr. at 500° C. *Curve 6* Hard-rolled, annealed 1 hr. at 500° C + 15⁰/₀ tensile overstrain + 1 hr. at 500° C

vibratory, and wave speed situations are summarized in the accompanying table for the smallest specimen diameter, 0.091 in. Measurements were made using nearly all the different types of experimental methods.

Specimen Preparation	E Avg. Static $\times 10^6$ psi	E Avg. Dynamic $\times 10^6$ psi	E Total Avg. $\times 10^6$ psi	μ Avg. Static $\times 10^6$ psi	μ Avg. Dynamic $\times 10^6$ psi	μ Total Avg. $\times 10^6$ psi
¹/₂ H	16.90	16.69	16.83	5.96	6.05	5.99
¹/₂ HT	18.54	17.76	18.23	6.86	6.69	6.79

The ratio of the total average E moduli is 1.082, and the ratio for the total average μ moduli is 1.132. A change of p from $p = 1$ for copper to $p = 0$ for the alloy would provide a ratio of 1.108. A similar shift in p for the alloy α-brass is noted later in the present chapter.

Two of the experiments described by RICHARDS (1952) are tensile and compression experiments in 1.8% Be-Cu, carried out with great accuracy by MILLER of the U.S. National Bureau of Standards, using Tuckerman optical gauges. Both of these experiments in the $^1/_2$ HT solid provided predicted stable moduli for the infinitesimal deformation up to what is referred to as the proportional limit, after which, as may be seen in Fig. 6.7, transitions from $s = 8$, $p = 1$ to $s = 9$, $p = 1$ are observed for tension and $s = 10$, $p = 1$ for compression.

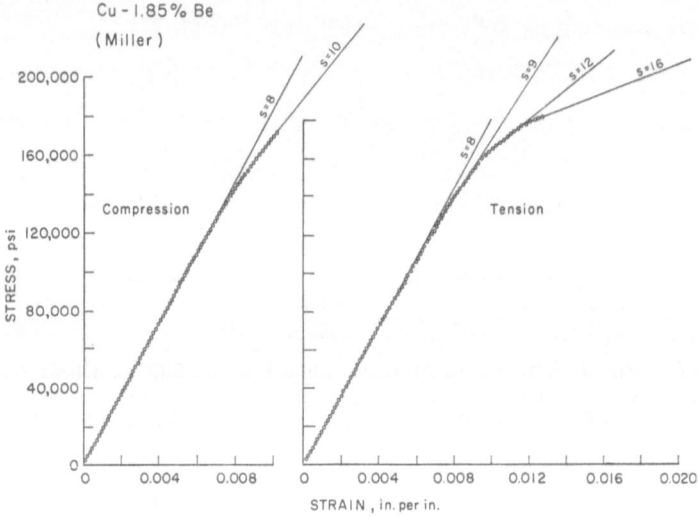

Fig. 6.7. Tuckerman optical gauge measurements of compression and tension E moduli, showing transitions at proportional limit

The study of the multiple elasticity phenomenon described immediately below has included a series of uniaxial stress experiments in iron, steel, aluminum, zinc, magnesium, copper, and α-brass in most of which repeated loadings in the small strain region were followed by large deformation measurements. A second series of experiments, to be described in Chapter VII, were performed to show that during very large deformation the rising portion of a Portevin - le Chatelier step was subject to slope variations in the form of the same discrete distribution when a transition in the mode index, r, occurred.

Multiple Elasticity Experiments

The primary purpose of this experimental series was to establish that multi-linear infinitesimal strain regions exist and that they have slopes which are numerically given in the form of the discrete distribution.

Since the experiments were conducted in uniaxial tension and compression, it is necessary to relate the measured E moduli to the zero-point isotropic linear elastic shear moduli distribution described in the previous chapter, knowing that for isotropic linear elastic solids, $E = 2\mu$ $(1 + \nu)$ here ν is Poisson's ratio. Then from Eqs. (5.1) and (5.4) we write Eq. (6.1).

$$E = \frac{\delta\sigma}{\delta\varepsilon} = 2\,(1 + \nu)\times 1.03\times(2/3)^{s/2 + p/4}A\,(1 - T/2T_m). \quad (6.1)$$

ALUMINUM POLYCRYSTALS , 99.16% Purity , 300°K

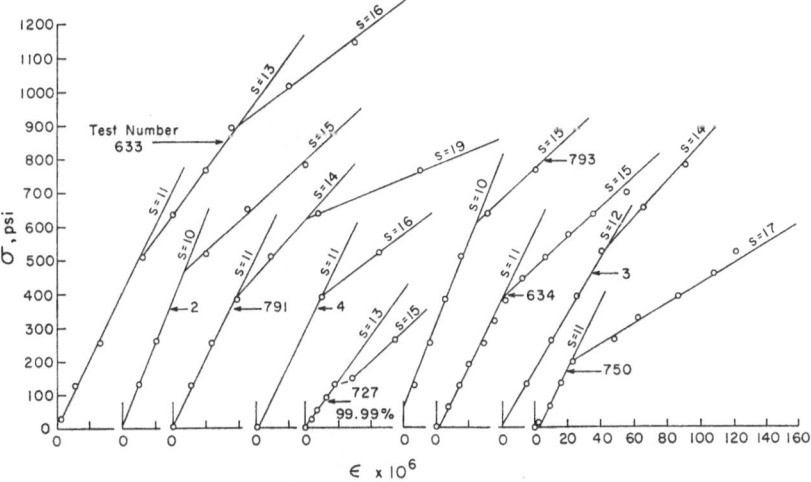

Fig. 6.8. Multiple E moduli determinations at small strain

In what follows, predicted values of the moduli will be identified through the values of the integers s and p which apply in the particular instance. Figs. 6.8 and 6.9 show the small strain portion of 18 polycrystalline aluminum experiments. Two of these experiments, Nos. 636 and 637, are for work-hardened aluminum, the remaining measurements (whose finite distortional deformation parabolas were described in Chapter III, Figs. 3.4 and 3.5) being for completely annealed aluminum. The circles in Figs. 6.8 and 6.9 represent the experimental data; the solid lines are theoretical slopes from Eq. (6.1).

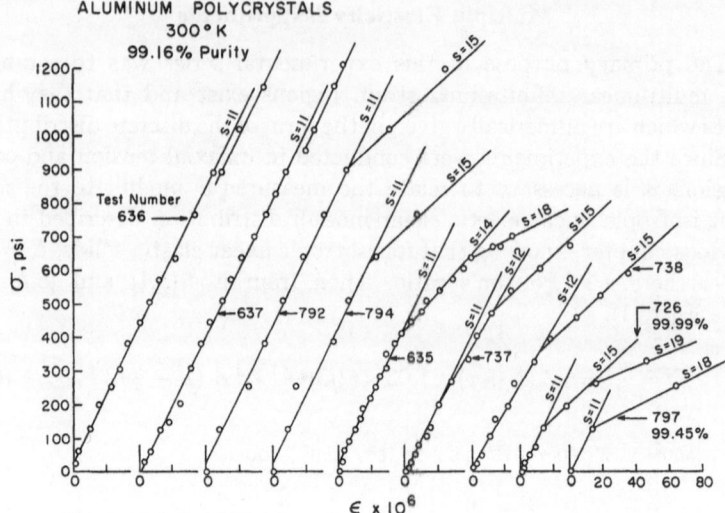

Fig. 6.9. Multiple E moduli determinations at small strain

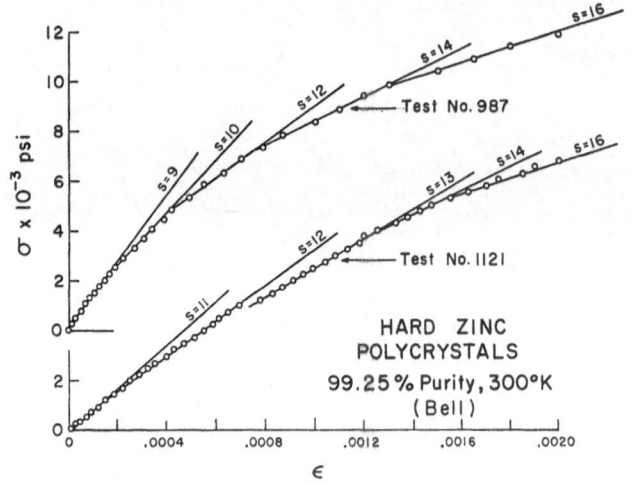

Fig. 6.10. Multiple E moduli at small strain

In all but 5 of these experiments, the initial modulus has the stable value for aluminum of $s = 11$, $p = 0$, as noted in Chapter V. In all experiments including these 5, beyond the first proportional limit the predicted multiple elasticities are observed.

A similar behavior is observed for the hard zinc polycrystals of Fig. 6.10 and for the annealed zinc polycrystals of Fig. 6.11. The stable initial modulus for Zn from Chapter V has a value of $s = 9$, $p = 0$

which is the value of the initial modulus shown for one of the two hard zinc polycrystals, and for 3 of the 6 annealed zinc polycrystals. Again one finds that initial and subsequent slopes follow the predicted multiple elastic behavior.

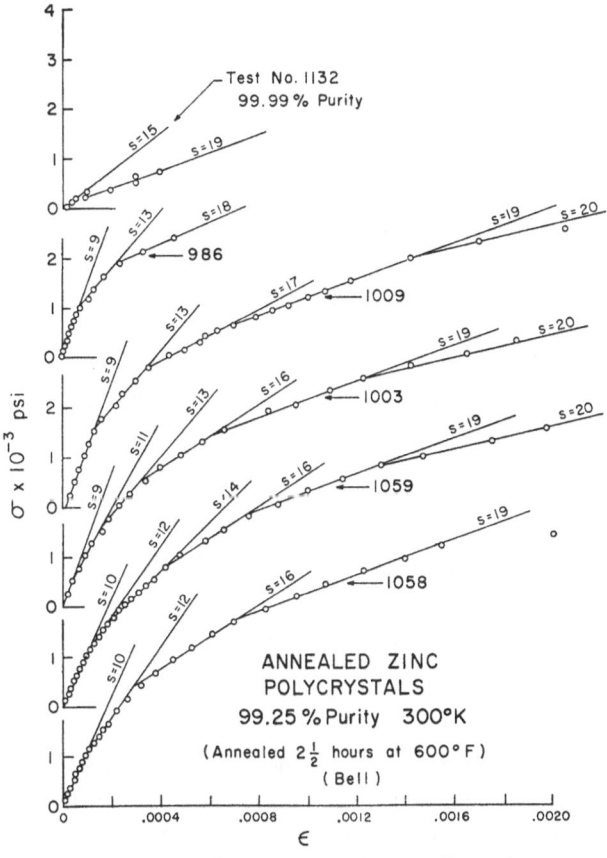

Fig. 6.11. Multiple E moduli at small strain

The parabola coefficients at large deformation for these annealed zinc specimens and for one of the hard zinc specimens, were shown in Fig. 3.10 of Chapter III.

The E modulus in iron is perhaps one of the best-known numbers in metals technology. The accepted value of $E = 20,900$ kg/mm² or $E = 29.7 \times 10^6$ psi, corresponds precisely to a value of $s = 6$, $p = 0$. As may be seen from Fig. 6.12 in 99.85% purity ingot iron annealed 48 hours at 1,640° F, the initial modulus is indeed the expected stable one of $s = 6$, $p = 0$. At a very small stress, however, a sharp change in slope occurs which is followed by a long straight region whose slope now has the value of the modulus $s = 9$, $p = 0$ to over 16,000 psi when discontinuous yield occurs at what is usually referred to as the elastic limit.

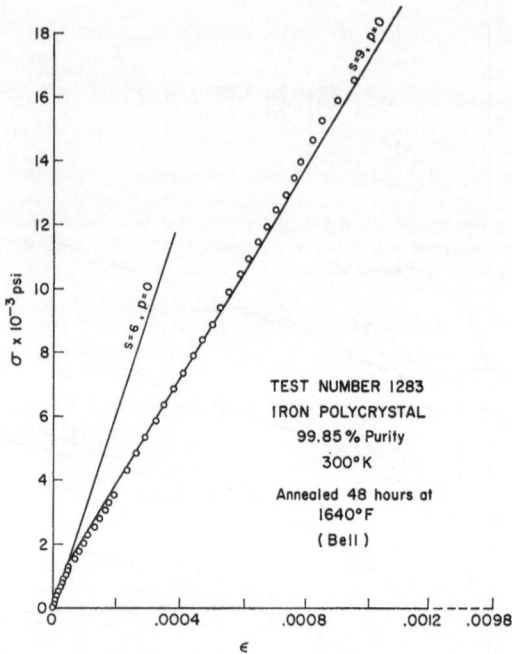

Fig. 6.12. Two moduli before discontinuous yield. The initial modulus for $s = 6$, $p = 0$ is $E = 29.7 \times 10^6$ psi; for $s = 9$, $p = 0$, $E = 16.2 \times 10^6$ psi

In Figs. 6.13 and 6.14 a series of 99.85% purity ingot iron polycrystals annealed one hour at 1,200° F and 24 hours at 1,640° F, together with one experiment in hard iron and two experiments in low carbon steel, provide the distribution of initial and subsequent moduli shown.

For all these experiments, $p = 0$. For one experiment the stable initial value of $s = 6$ is observed. For all the rest, the initial value is either $s = 7$, $s = 8$, or $s = 9$, with discrete changes in slope to other predicted values up to the point of discontinuous yield whose stress of initiation and final strain has been designated in each instance.

These high purity iron compression experiments were continued on to failure after the discontinuous yield had occurred. As may be seen from the σ^2 vs ε plots of Fig. 6.15, the large deformation is parabolic, with predicted finite deformation mode indices, r. These data were not shown earlier in Chapter III and hence provide additional experimental evidence that the finite distortional deformation of iron is in agreement with the writer's linear temperature dependent generalized parabolic stress-strain function.

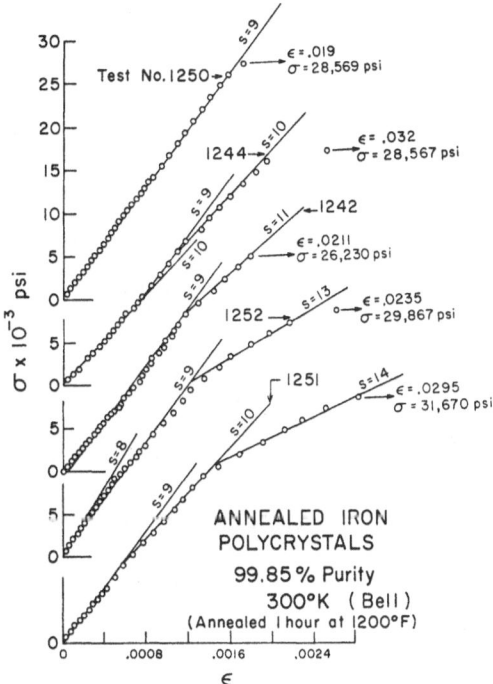

Fig. 6.13. E moduli distributions in iron. Arrows designate discontinuous yield whose stress and strain is specified for each experiment

The small strain portion of uniaxial experiments in nickel, copper, and magnesium are shown in Fig. 6.16. The stable values for nickel are $s = 6$, $p = 0$; for copper, $s = 8$, $p = 1$; and for magnesium, $s = 13$, $p = 0$.

The initial moduli for these fully annealed specimens of Fig. 6.16 differ from these stable values. For the nickel experiment of FILBEY (1965) the initial modulus is $s = 8$, $p = 0$, changing at a strain of around 80×10^{-6} to the predicted value of $s = 6$, $p = 0$. For two of the copper tests the initial moduli are $s = 10$, $p = 1$, and for the third, the initial modulus is $s = 12$, $p = 1$. For both of the annealed magnesium experiments the initial moduli all the way to discontinuous yield are $s = 15$, $p = 0$ (i.e., $E = 4.32 \times 10^6$ psi) instead of the stable value of $s = 13$, $p = 0$ (i.e., $E = 6.48 \times 10^6$ psi). For each of these polycrystalline solids, longitudinal wave propagation E moduli determinations in long bars

11*

have provided stable moduli. For example, 8 such wave propagation experiments in 6 ft. bars of the magnesium rod from which the annealed specimens were produced gave an average value of 6.38×10^6 psi (i.e., $s = 13$, $p = 0$), with the highest observed experimental value of 6.56×10^6 psi and the lowest, 6.17×10^6 psi.

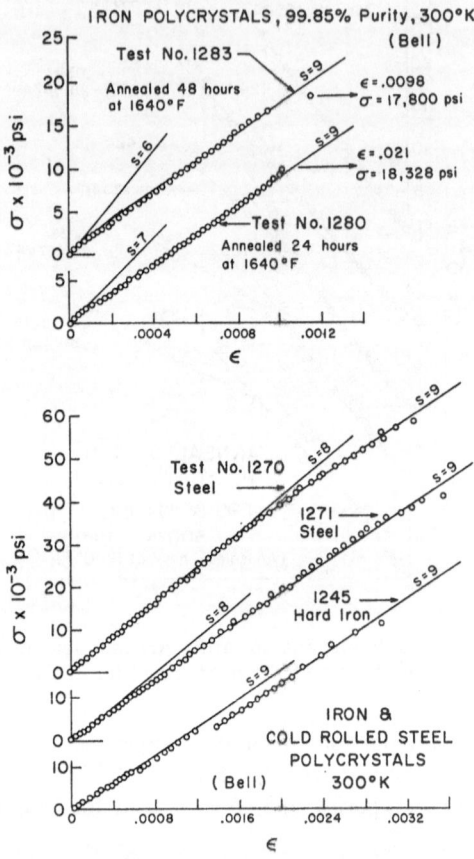

Fig. 6.14. Multiple E moduli for annealed iron, hard iron, and cold-rolled steel before discontinuous yield

The large deformation portion of FILBEY's (1965) nickel experiment was shown in Fig. 3.20 in Chapter III. The large deformation of the three copper experiments is shown in Fig. 6.21 below. The large deformation portion of the annealed magnesium experiments is shown in Fig. 6.17.

No quasi-static magnesium data were shown in Chapter III, but both quasi-static and dynamic data have been described in Chapter II. The fact that the upward-turning magnesium data of Fig. 6.17 are parabolic further extends the present finite distortional deformation generalization. One notes upward transitions from $r = -1$ to $r = -2$ in each instance. The dynamic data in this solid, described in Chapter II, had an initial parabola coefficient of $r = 2$, instead of the $r = -1$ of Fig. 6.17.

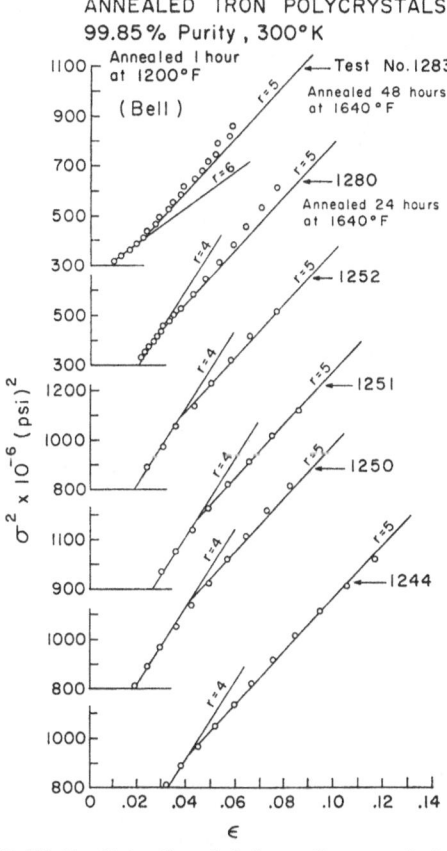

Fig. 6.15. Finite distortional deformation parabolas in iron

The most important multiple moduli experiments with respect to the finite deformation behavior are those in which a series of reversed loadings at small strains were followed by loading the specimens in finite deformation. In these experiments the reversed loading could be made to occur either before or after one or two Portevin - le Chatelier discontinuous yieldings had taken place.

The multiple elastic phenomenon can be accentuated by varying the prior thermal history. One group of specimens (1259, 1260, 1261,

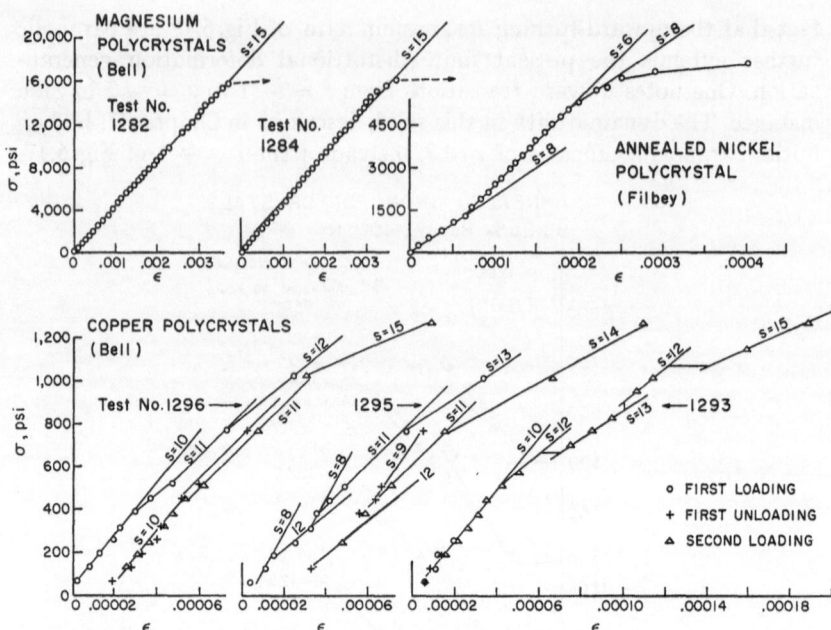

Fig. 6.16. Multiple E moduli in magnesium, nickel, and copper. Note the multiple loading of copper experiments

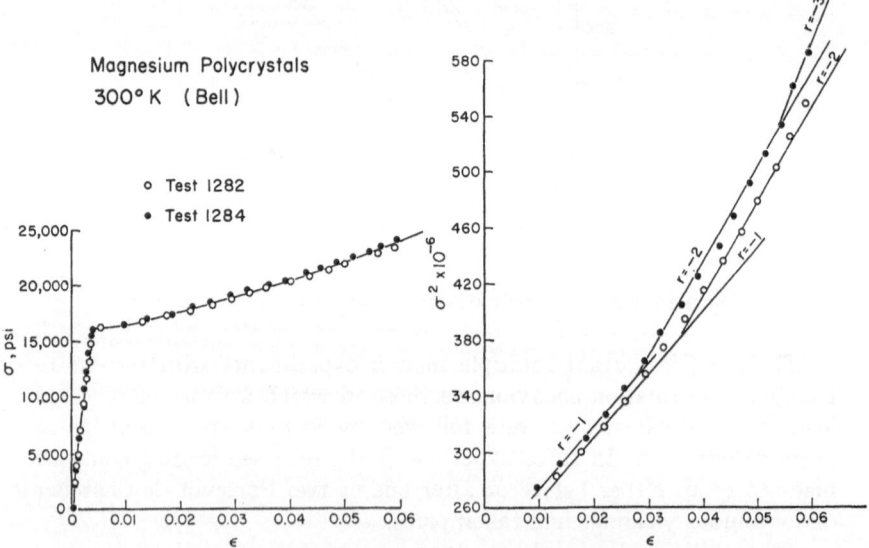

Fig. 6.17. Finite distortional deformation for magnesium polycrystals

1262 and 1263) were held just below the melting point of 1,220° F for 10 minutes; the temperature was then lowered to the usual 2 hour annealing at 1,100° F; and, finally, the specimens were furnace cooled. Following deformation these specimens were etched to check grain size. Specimen 1259 had the conventional fine grain of the usual 1,100° F anneal, but the other 4 specimens were seen to have grown one or two large central grains at each end of the specimen. The aluminum specimens of experiments 1264 through 1269, and the square specimens, 1306 and 1312, were annealed in the usual manner at 1,100° F for 2 hours and furnace cooled, and in every instance etching showed the usual fine-grained structure throughout the specimen. All but the two square specimens were 3 in. long and 1 in. in diameter. All measurements were in uniaxial compression.

Experiments Nos. 1260 and 1259 were not carried beyond the small strain region so that these specimens might be examined post-deformation with minimum permanent set. Experiment 1260 included three cycles of loading and unloading. The maximum stress was kept sufficiently small so that no plastic deformation was introduced until the third loading cycle. Reversible transitions from $s = 10$, $p = 0$ to $s = 13$, $p = 0$ are seen in the first two loadings. The final unloading modulus of the third cycle after two Portevin-le Chatelier steps had occurred was $s = 13$, $p = 0$. The permanent deformation was 20×10^{-6}. The results of experiment 1260 are shown in Fig. 6.18. It should be remembered that the stable modulus for aluminum is $s = 11$, $p = 0$.

In experiment No. 1259, shown in Fig. 6.19, a Portevin-le Chatelier step at approximately 450 psi occurred during the first loading cycle and, again, the unloading slope was that of an $s = 13$, $p = 0$ modulus. The modulus during the rising portion of the Portevin-le Chatelier step in this cycle and in the Portevin-le Chatelier steps which occurred in the second cycle, are $s = 14$, $p = 0$, in common with most of the other aluminum polycrystalline experiments considered below. The second unloading and the third loading have a reversible change of slope from $s = 14$, $p = 0$ to $s = 16$, $p = 0$.

Also shown as dashed lines in Fig. 6.19 are the calculated slopes for the writer's parabolic law for $r = 3$.

In Fig. 6.20 is shown the small strain portion of experiment 1261. No Portevin-le Chatelier steps occurred during the first loading cycle, but three steps beginning slightly above 400 psi were permitted during the second loading cycle. For all three cycles in this experiment the slope is that of an $s = 12$, $p = 0$ modulus. Following the third loading in which a Portevin-le Chatelier step to an $s = 14$, $p = 0$ modulus is observed, the experiment was carried on into the region of finite deformation which began as an $r = 3$ parabola with a transition at low strain to an $r = 4$ parabola, as may be seen in the σ^2 vs ε plot for this experiment shown in Fig. 6.21.

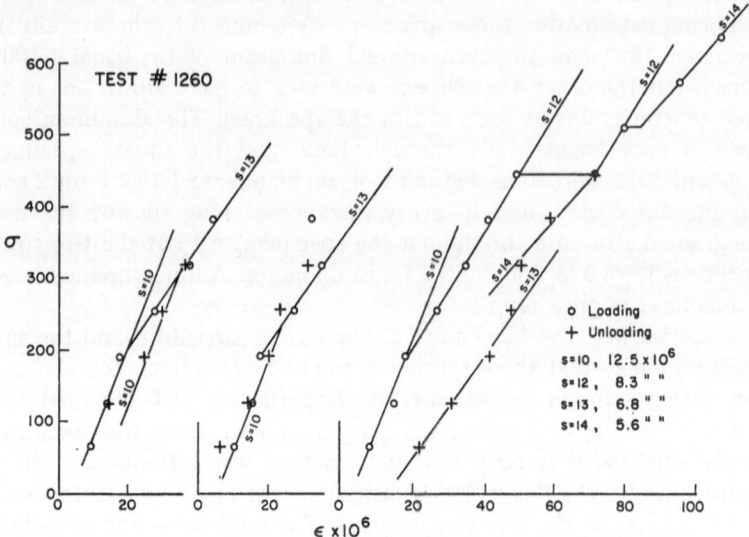

Fig. 6.18. Three loading cycles at small strain, with plastic deformation on third cycle

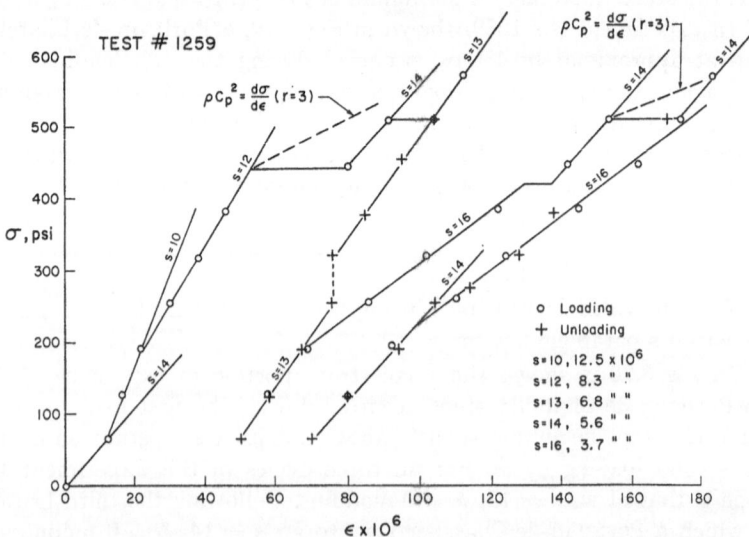

Fig. 6.19. Three loading cycles at small strain, with plastic deformation on first cycle

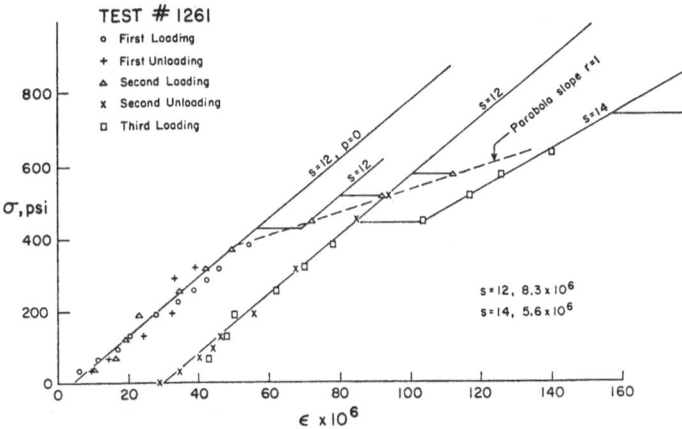

Fig. 6.20. Three loading cycles at small strain, with plastic deformation on second cycle

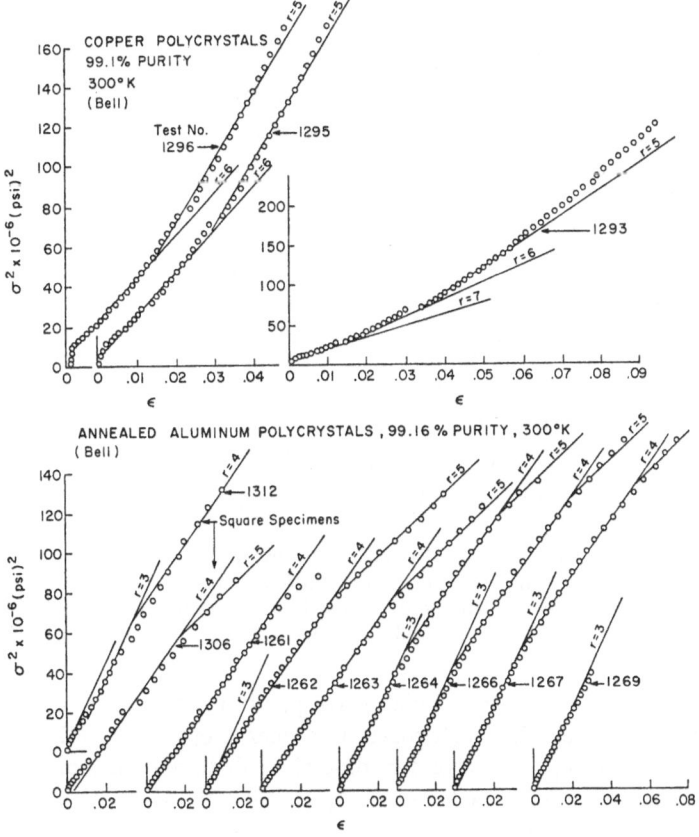

Fig. 6.21. Finite deformation of copper and aluminum following multiple small strain loading cycles. Note higher mode indices, r, when comparing with polycrystalline data of Chapter III

The common final modulus of $s = 14$, $p = 0$ of experiments 1259, 1260, and 1261 provide a correlation between the finite deformation mode index, r, and the moduli index, s. The $s = 14$ differs from the

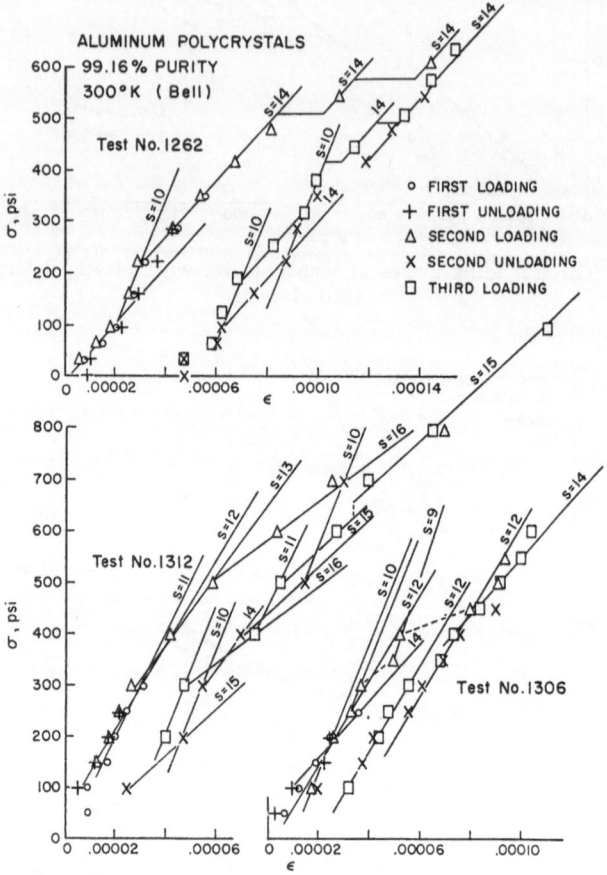

Fig. 6.22 (a) and (b). Reversed cyclical loading experiments at small strain, showing multiple E moduli. Specimens 1306 and 1312 have 1.80 cm per side, square cross-section; other specimens have circular cross-sections with 2.50 cm diameters

stable value $s = 11$ by 3, which corresponds to the $r = 3$ of the finite deformation. For the Portevin-le Chatelier steps in experiment 1261 of Fig. 6.20 during the second loading the calculated parabola slope for $r = 1$ (shown as a dashed line) is consistent with the unit change in s from 11 to 12. As might be expected, not all experiments provide this correlation. Plastic deformation usually begins following such a small elastic slope change and, as is characteristic of finite deformation mode transitions, the initial mode index r is not predictable in a particular experiment.

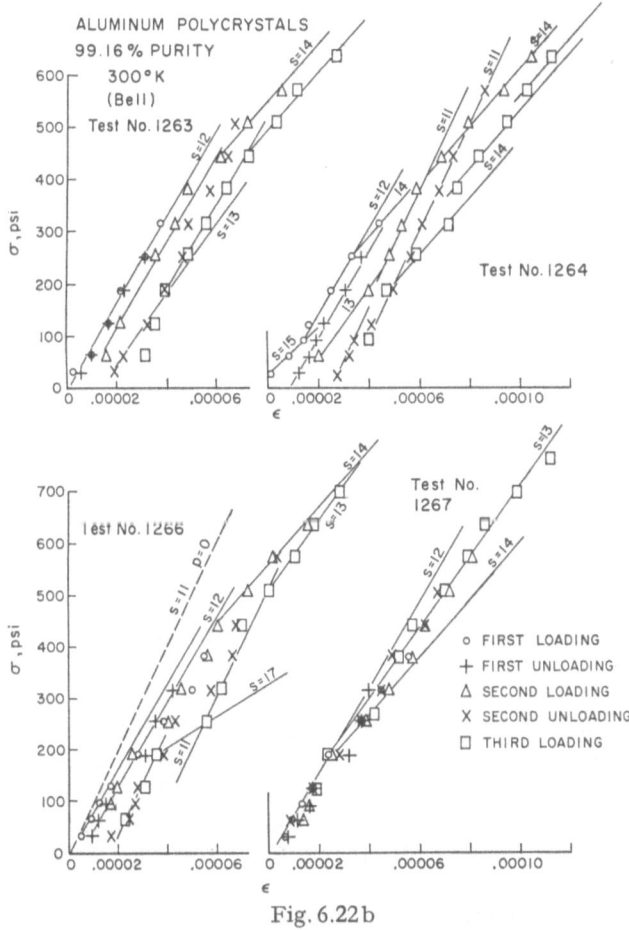

Fig. 6.22 b

The small strain multiple loading behavior of the designated completely annealed aluminum polycrystalline experiments is shown in Figs. 6.22 (a) and 6.22 (b). For all of these experiments, including the two square cross-section experiments Nos. 1306 and 1312, the finite deformation parabolas were obtained, as may be seen in Fig. 6.21.

These two square cross-section uniaxial compression experiments were performed following a suggestion of M. BEATTY of the need to demonstrate that the observed instabilities in moduli did not have a geometric origin.

When cyclical small strain loading is introduced, large deformation parabolas have initial mode indices of $r = 3$ or $r = 4$ with the predominant large strain mode of $r = 4$. Except for one group of experiments whose prior metallurgical treatment and whose end constraints had been deliberately varied, these stress-strain curves are in general lower than those shown in Fig. 3.4 of Chapter III.

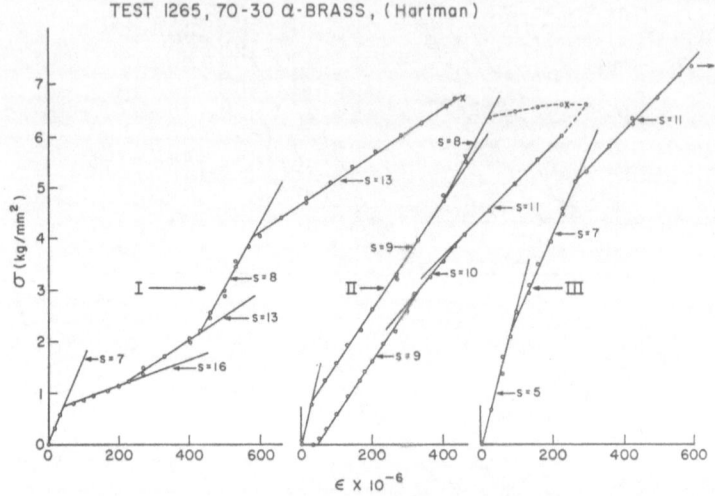

Fig. 6.23. Three loading cycles in annealed 70—30 α-brass, showing numerous E moduli

HARTMAN (1967) performed a small strain experiment in 70—30 α-brass, with three loading cycles followed by large strain measurement. For α-brass the stable modulus is $s = 9$, $p = 0$. The multiplicity of values which HARTMAN obtained in this annealed polycrystalline solid are indicated in Fig. 6.23. HARTMAN has shown that the large strain portion of this experiment gave parabola coefficients in very close agreement with the present generalization.

One thus sees from the experiments described in this chapter that multiple linear elastic regions do exist in each of the several solids which have been considered here, and their slopes do have the predicted distribution of Eq. (6.1).

The writer's conjecture that the multiple elasticities of an *individual* element quantitatively occur in precisely the same discrete distribution as that *among* the elements, is firmly established.

The density of multiple elasticities is increased by varying the prior thermal history, or by introducing small strain cyclical loading in conventionally annealed solids. Stable isotropic elastic moduli are observed in solids which meet prescribed metallurgical specifications of hardness, heat treatment, etc. In quasi-static testing, such solids do not exhibit multiple elasticities until relatively large stress, when the proportional limit is reached. Dynamic measurements in either work-hardened or annealed solids usually provide stable moduli. This is particularly true of ultrasonic measurements which are made at near zero stress, and ultrasonic measurement is the most frequently used experimental technique for the determination of elastic constants at the present time. The multiple elastic experiments of this chapter, of course, are all dead-weight quasi-static uniaxial measurements.

The important point is that when conditions are introduced which proliferate the multiple elasticities in an experiment, the slopes observed fall into the discrete distribution pattern of the zero-point parabola coefficients and the zero-point isotropic linear elastic moduli distribution among the elements.

One must conclude on the basis of experiment itself that linear elasticities are independent of the particular crystal structure and that all values, including stable moduli, designate members of a series of states common to all of the nearly 60 crystalline solids for which infinitesimal deformation data is known. The probabilistic aspect of this phenomenon obviously is of some interest in the field of linear elasticity. That it is even more significant with respect to the finite distortional deformation generalization developed in this monograph, will be discussed in Chapter X.

Chapter VII

The Portevin - le Chatelier Effect

At sufficiently low average strain rates in dead-weight quasi-static uniaxial stress experiments instabilities appear in the form of a series of successive steps superimposed upon the basic parabola of the large deformation. The first serious study of this phenomenon, in 1923, was that by PORTEVIN and LE CHATELIER, after whom the phenomenon is now named. An example of this Portevin - le Chatelier effect may be seen in Fig. 7.1 in the present writer's experiment 1219 for a 99.16% aluminum polycrystal. This experiment, shown in a σ^2 vs ε plot in Fig. 3.3, is a dead-weight uniaxial tension measurement at a constant stress rate of 4.5 psi sec^{-1}.

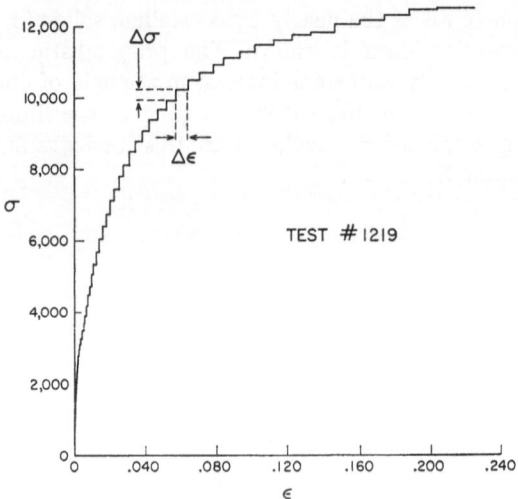

Fig. 7.1. Typical Portevin - le Chatelier effect constant stress rate tension experiment in annealed polycrystalline aluminum

The vertical portion of the steps $\Delta\sigma$ in such a slow dead-weight measurement rises in a relatively long time interval with slopes in the range of linear elastic moduli. In general the horizontal portion of the steps $\Delta\varepsilon$ occurs very rapidly although slow moving instabilities have been observed at low strain.

That this Portevin - le Chatelier effect is related to the finite deformation mode and transition structure of the present large deformation generalization is demonstrated by two experimental facts. First, the vertical portion of the Portevin - le Chatelier step $\Delta\sigma$ always rises from, and the horizontal portion of the step $\Delta\varepsilon$ always returns to, the governing parabolic stress-strain function irrespective of the mode index r, which may be operative in the region of deformation examined. Second, marked changes in the incremental stress and incremental strain behavior are observed to occur at transition strains whether or not a change in the mode index, r, of the governing stress-strain law actually occurs. This behavior characterizes both the polycrystalline and the single crystal stage III deformation of nearly all of the dead-weight aluminum experiments, whether compression or tension, described in Chapters III and IV.

HARTMAN (1967) has shown that the Portevin - le Chatelier effect in α-brass is similarly characterized by major changes in step behavior at the same transition strains found in aluminum. An inspection of data in Chapters III and IV as, for example, Figs. 3.6, 3.7, 4.12, 4.13, or 4.20 will reveal that the writer has shown that the Portevin - le Chatelier phenomenon may occur for all impurities considered, from $99.16^0/_0$ to $99.99+^0/_0$ in aluminum in the temperature range from $273°$ K to $423°$ K. Among the proposed explanations of the Portevin - le Chatelier effect have been the solute atom diffusion-vacancy hypothesis of COTTRELL (1953), the negative strain rate COULOMB friction "stopped motion" analogue of BODNER and ROSEN (1967), and the particle wave hypothesis of FITZGERALD (1966). It has been shown by ERICKSEN (1963) in his non-linear continuum theory of oriented solids that such a phenomenon as the Portevin - le Chatelier effect may have a natural origin in the macroscopic deformation itself, without reference to atomistic mechanisms. In 1949 MCREYNOLDS carried out a series of uniaxial stress, constant stress rate experiments on an experimental apparatus nearly identical with that used 16 years later by SHARPE (1966a, b, c) and by the present writer. In such experiments careful attention must be paid to the fineness of scale in determining whether or not the phenomenon is visible. One not only must be able to measure extremely small steps in some circumstances but also one must use stress increments fractionally smaller than the height of step being observed. MCREYNOLDS discovered that the finite deformation was inhomogeneous in that the individual step, when it occurred, propagated through the solid at extremely low velocities; i.e., from a few millimeters per second to around 100 centimeters per second.

In 1963 DILLON independently rediscovered the Portevin - le Chatelier effect while studying the thermodynamics of cyclical plastic torsional deformation in annealed aluminum and copper. DILLON has made by far the most extensive study of the Portevin-le Chatelier effect to date. His torsion measurements, like those of McREYNOLDS (1949) in tension and those of BELL and STEIN (1962) in compression, consisted of local observations over one or more small regions of the deforming specimen. The more recent experiments of SHARPE (1966a, b, c) and of BELL have considered averaged behavior over long gauge lengths in order to obtain reproducibility. DILLON's (1963, 1966) studies of the Portevin-le Chatelier effect included a comparison of experiments requiring a few hours, to very slow dead-weight loading torsion experiments covering a period of many weeks. DILLON (1966) and KENIG and DILLON (1966) also extensively studied the McReynolds slow waves and proposed a general explanation of plastic deformation including finite amplitude wave propagation based on the Portevin-le Chatelier effect as a fundamental macroscopic mechanism of distortional deformation in continuum mechanics.

BELL and STEIN in 1962 showed that the Portevin-le Chatelier effect provided a mechanism for explaining the elastic velocity of an incremental loading wave in a prestressed plastic field. That incremental loading waves traveled with linear elastic velocity was first observed by BELL in 1949 and described in a report in 1951. BELL and STEIN (1962) showed that incremental loading waves traveled at the linear elastic bar velocity only if a dynamic prestress had reached a static plateau behind the finite amplitude wave front; i.e., the dynamically produced fixed stress plateau behind the wave front was at the bottom of a Portevin-le Chatelier step. Incremental loading waves which propagated into a *dynamically increasing* prestressed wave front did not propagate with a linear elastic velocity. Furthermore, when the incremental loading wave had an amplitude higher than that expected for the stress increment $\Delta \sigma$ of a Portevin-le Chatelier step, only the initial tit of the wave propagated with the linear elastic bar velocity, the remainder of the increment being propagated at slower plastic wave speeds. From such incremental wave studies it has been established that conditions required to produce a Portevin-le Chatelier step occur in a measured time interval as short as 10×10^{-6} secs.

In the dead-weight compression of a 99.99% purity single crystal experiment of the writer (see Fig. 4.13) which was previously shown to agree with a predicted $r = 6$ parabola, the strain-time behavior was continuously monitored for 15 days (357 hours). Continuously monitored strain stability at the foot of a Portevin-le Chatelier step was observed to be maintained for a period as long as 94 hours. Using the same apparatus in a 1962 dead-weight uniaxial compression experiment in polycrystalline aluminum, Fig. 4.12, the writer observed stability maintained for a period of 17 hours at a stress of 0.0067 kg/mm² below the stress necessary to trigger a horizontal $\Delta \varepsilon$ increase. This 0.0067 kg/mm² was but a small fraction of the total height of the step being measured. Thus, one observes that stability at the foot of a step may be established in microseconds and maintained for a period of several days.

In a dead-weight loading experiment in which small increments of stress are added at repeated intervals, the appearance or non-appearance of the Portevin-le Chatelier effect is dependent upon the magnitude of the increment which is added. For a given increment of loading the appearance or non-appearance of the phenomenon is dependent upon the amount of time which is allowed to elapse between increments, provided the increment is sufficiently small so that the Portevin - le Chatelier effect may be observed. This may be seen in the 357 hour test of Fig. 4.13. In this dead-weight compression experiment on a 99.99% aluminum single crystal at 300° K, 83 load increments of 0.0067 kg/mm² were applied at various times during the experiment. The average strain rate was $\dot{\varepsilon}$ avg. $= 6 \times 10^{-9}$ for the total experiment. During the 7th day, 16 such increments were added uniformly over a period of 7 hours producing an average strain rate during this time of $\dot{\varepsilon} = 5.5 \times 10^{-8}$ sec^{-1}. During this time the scale of the Portevin-le Chatelier phenomenon was either too small to be observed or non-existent (the interval of strain for these measurements is between $\varepsilon = 0.0027$ and $\varepsilon = 0.0042$). Following the addition of the last increment of this series of increments, at a strain of 0.0042 a continuously monitored interval of 16 hours was allowed to elapse without additional stress being added. At the end of this interval, subsequent loading of a series of 0.0067 kg/mm² increments produced a vertical $\Delta \sigma$ step.

When this Portevin-le Chatelier step had occurred the prior loading process was repeated with eleven 0.0067 kg/mm² increments being added during the next 8 hours, followed by a 94 hour stable strain constant stress period after the last increment. Again a vertical step of 3 increments was observed. This time, however, the vertical step was followed by a conventional series of observable Portevin-le Chatelier steps not requiring a long delay for their inauguration. A second 16 hour constant stress interval was introduced at a strain of 0.006 which again produced a vertical step which was followed by 3 conventional Portevin-le Chatelier steps not requiring a long delay. To be certain that the Portevin-le Chatelier phenomenon had been established in a 99.99% aluminum single crystal without the necessity for introducing the longtime trigger mechanism, a final constant stress interval of 23 hours was introduced at the foot of a step at $\varepsilon = 0.0075$. During this time strain stability was observed.

What is of major importance in this experiment, in addition to the production of the Portevin - le Chatelier effect at low strain in a high purity single crystal, is that despite the varied loading history during the period of 15 days, the ratio $\frac{\Delta\sigma}{\Delta\varepsilon}$ was such that the deformation returned to the predicted parabola $r = 6$. This, of course, is the situation in the more uniform constant stress rate experiments, and in the interrupted compression polycrystalline measurement described in Chapter III, Figs. 3.6 and 3.7, and Chapter IV, Figs. 4.12, 4.13, and 4.20.

All 13 of the constant stress rate experiments of Fig. 3.3 and experiments 786, 787, 788, 914, 971, 1204, and 1207 of Fig. 3.4, were performed for the purpose of studying the Portevin - le Chatelier effect in low purity aluminum polycrystals. All of these experiments, as well as the 16 single crystal experiments of Fig. 4.6 and two bi-crystal experiments of Figs. 4.18 and 4.19, were performed in the writer's laboratory. The 700 and 900 series Portevin - le Chatelier experiments, including the single crystal data of Fig. 4.6, were measurements of SHARPE (1966 a, b, c) in his study of the proliferation of the Portevin - le Chatelier effect in the presence of grain boundaries. The 1200 series Portevin - le Chatelier experiments of the present writer were carried out for the purpose of studying the role of multiple elasticities and maximum straining rates in the Portevin-le Chatelier effect.

Unlike the earlier experiments of McREYNOLDS (1949), BELL and STEIN (1962), DILLON (1963), and BELL (1962; see Figs. 4.12 and 4.13) all the experiments mentioned in the previous paragraph were performed with the strain being integrated over a 2 in. gauge length. The inhomogeneity of the deformation as exhibited in McREYNOLDS' slow waves and as extensively studied by DILLON, were by this integration statistically smoothed, providing a reproducibility in the Portevin-le Chatelier phenomenon from one experiment to another. Because of this reproducibility, SHARPE (1966a, b, c), through a comparison of the Portevin - le Chatelier phenomenon in single crystals, bi-crystals, and polycrystals of an identical low purity, was able to show that the presence of grain boundaries did proliferate the step density and detail. SHARPE also showed that the first inauguration of Portevin - le Chatelier steps in single crystals and polycrystals occurred at stress and strain values given by the ratios of the TAYLOR (1938) aggregate theory, Eq. (4.8).

The underlying parabolicity of the constant stress rate Portevin-le Chatelier experiments was shown in Figs. 3.3 and 3.4, demonstrating that the $\Delta\sigma/\Delta\varepsilon$ ratio of the stepped structure is superimposed upon the generalized parabolic stress-strain function. For the 1200 series experiments in these constant stress rate uniaxial tension data, plotting the stress increment $\Delta\sigma$ as a function of strain, and the strain increment $\Delta\varepsilon$ as a function of strain, provides an interesting interrelationship between this phenomenon and the transition strains of the finite deformation mode and transition structure. Such data for 8 experiments in the 1200 series are shown in Fig. 7.2.

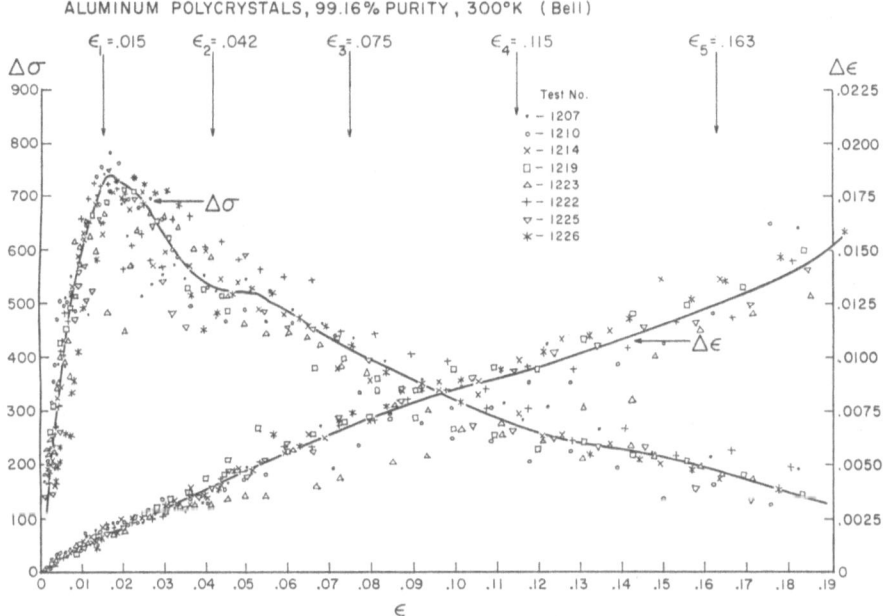

Fig. 7.2. Portevin - le Chatelier effect stress increments and strain increments plotted against the average strain for several constant stress rate experiments in polycrystalline aluminum. Note the changes in behavior at the critical strains (arrows)

The major peak in the vicinity of the first transition strain of $\varepsilon_N = 1.5\%$ can be noted in the data of SHARPE (1966a, b, c). This is again exhibited as a maximum peak in the $\Delta\sigma$ vs ε averaged curve of the 1200 series data of Fig. 7.2 and as a slope discontinuity in the linear $\Delta\varepsilon$ vs ε curve. Some of the 1200 series experiments provided a second, lower peak in the $\Delta\sigma$ vs ε curve at the second polycrystalline transition strain, $\varepsilon_N = 4.2\%$.

An examination of the data in Figs. 3.3 and 3.4 reveals that whereas one of the experiments of Fig. 7.2, namely 1204, underwent a change in mode index from $r = 2$ to $r = 3$ at the first transition strain of $\varepsilon_N = 1.5\%$; all of the remaining 7 experiments underwent no change of mode index until the second transition strain at $\varepsilon_N = 4.2\%$, where a transition from the mode index $r = 3$ to $r = 5$ occurred. This sensitivity of the Portevin - le Chatelier effect detail at a transition strain at

12*

which no change of mode index occurs obviously will have to be considered in developing an understanding of the mode and transition structure itself.

In SHARPE's low purity single crystal data (Chapter IV, Fig. 4.6) in all of which the Portevin - le Chatelier effect was observed, he noted that after deformation proceeded there were regions of strain in which the steps became experimentally non-observable. Since SHARPE (1966a, b, c) has described this behavior in great detail, it is only necessary here to summarize the results in Fig. 7.3. The transition strains for the resolved single crystal were given in Chapter IV; i.e., 4.6, 12.9, 23, 35, 50.5, and 82.6%. One notes in Fig. 7.3 that the beginning and ending of the regions of the Portevin - le Chatelier steps for these single crystals coincides with the transition strains. The first finite deformation mode transition from $r = 2$ to $r = 4$ occurred at the third single crystal strain of $\gamma_N = 0.23$ (see Fig. 4.6) where the second group of steps is observed to begin.

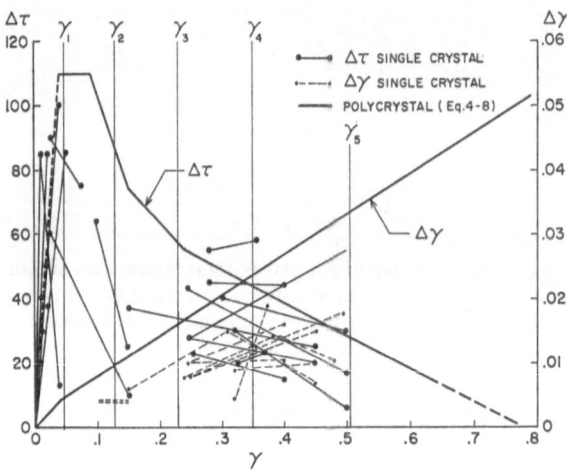

Fig. 7.3. The regions of occurrence of the Portevin - le Chatelier effect (SHARPE, 1966a and b) in the 99.16% purity aluminum single crystal data of Fig. 4.6, compared with the distribution of critical strains and with prediction from polycrystalline data through the aggregate theory ratios, Eq. (4.8)

DILLON (1963) has provided a detailed study of two torsion experiments in dead annealed 99.16% purity aluminum, which are described in Figs. 7.4 and 7.5 as Dillon test No. 613, and as "Dillon moderately slow torsion test". Test No. 613 was carried out over a period of three weeks. The S^2 vs s plots of Figs. 7.4 and 7.5 reveal that these torsion experiments also have a Portevin - le Chatelier effect superimposed upon an underlying parabola.

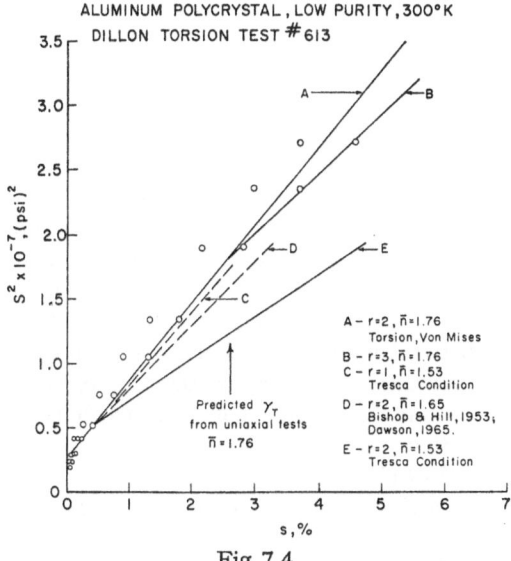

Fig. 7.4

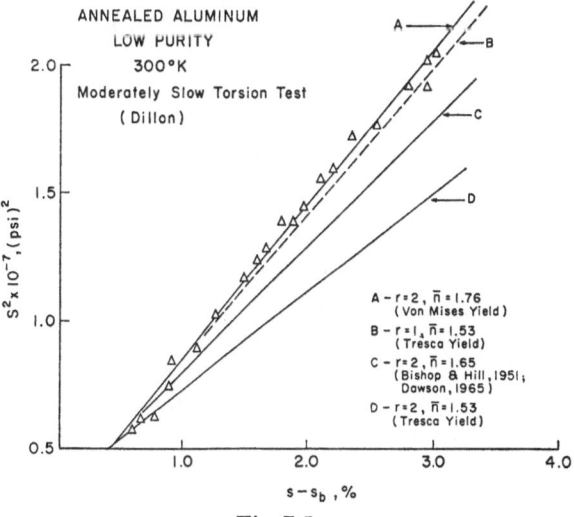

Fig. 7.5

BISHOP and HILL (1951) considered the applicability of the aggregate theory to the torsion of hollow tubes of the type considered in DILLON's experiment. By restating Eq. (4.8) for the aggregate theory ratios one obtains for the torsion of hollow tubes, Eq. (7.1):

$$S/\tau = \bar{n} = \gamma/s. \tag{7.1}$$

BISHOP and HILL suggested three possible values for $\bar{n}$: $\bar{n} = 1.76$, based upon a von Mises yield criterion; $\bar{n} = 1.53$ based upon a Tresca yield criterion; and $\bar{n} = 1.65$, based upon the aggregate theory analysis introduced by BISHOP and HILL in that paper (1951). DAWSON in 1965 included rotations in the aggregate theory analysis, and also obtained $\bar{n} = 1.65$ for the torsion of hollow tubes.

The predominant initial parabola coefficient for dead annealed fine-grained polycrystalline commercial purity aluminum has a deformation mode index of $r = 2$. Applying the aggregate theory ratios for uniaxial stress, Eq. (4.8), for this deformation mode index at room temperature, and then from the single crystal parabola thus determined calculating predicted torsional parabolas from the aggregate theory ratios of Eq. (7.1) for each of the three values of $\bar{n}$, one obtains the results shown in Figs. 7.4 and 7.5. These comparisons demonstrate that the operable ratio is that for $\bar{n} = 1.76$ of the von Mises yield criterion.

In Dillon test No. 613 a transition is observed from $r = 2$ to $r = 3$. The strain at which this transition occurs, $s_N = 2.6\%$, also is directly determinable from the first uniaxial transition strain, $\varepsilon_N = 1.5\%$ by means of the von Mises criterion ratio, $\bar{n} = 1.76$. In uniaxial stress data, as was shown by SHARPE, and as may be seen here in Fig. 7.2, the strain increment $\Delta\varepsilon$ is proportional to the strain, Eq. (7.2).

$$\Delta\varepsilon = K\varepsilon + C \qquad\qquad (7.2)$$

Two values of K are observed in aluminum, for regions below and above the first transition strain of $\varepsilon_N = 1.5\%$. HARTMAN (1967) has shown that Eq. (7.2) also is applicable to Portevin - le Chatelier data in α-brass, as will be shown below. That this also characterizes DILLON's torsion experiments in hollow tubes may be seen in Fig. 7.6 when Δs is plotted against s.

Fig. 7.6 also contains a plot of ΔS vs s for the torsional stress increment, which may be compared with the uniaxial stress behavior of Fig. 7.2.

Two experiments were carried out for the writer by C. JEFFUS in relatively high purity aluminum single crystals at 323° K and 373° K, to determine whether or not the Portevin - le Chatelier effect persisted as the temperature increased. Both of these experiments were performed in single crystals of 1 in. diameter, 3 in. long, as incremental dead-weight compression measurements. The data of Figs. 7.7 and 7.8 are the actually measured axial stress and strain. Since x-ray diffraction measurements of initial orientations were made, predicted deformation from resolved parabolas (dashed line) from Eq. (4.19) could be made.

Not only is the Portevin - le Chatelier effect dominant in these high purity single crystal data, but, as in all other experiments in which

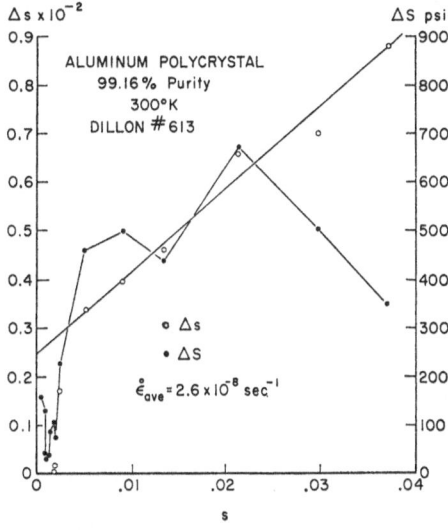

Fig. 7.6

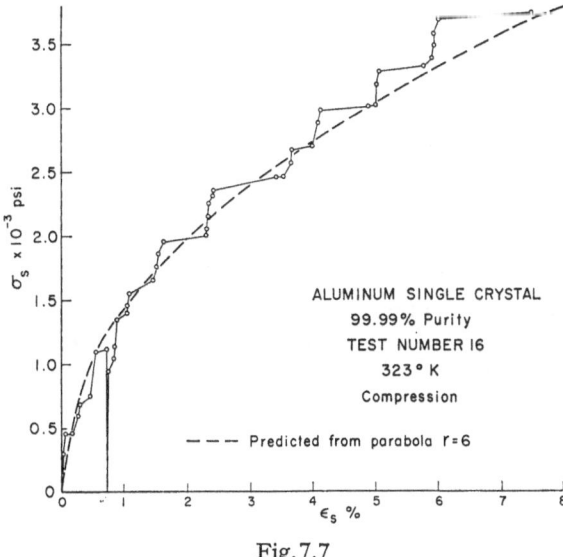

Fig. 7.7

this phenomenon is present, the steps are superimposed upon parabolas predicted from the present generalization. At 323° K the predicted parabola has a mode index of $r = 6$. With increasing temperature to 373° K, the deformation mode index has become $r = 7$.

One further aspect of the Portevin-le Chatelier effect which has not been described in earlier studies and which also is related to the mode and transition structure being described in this monograph, is the study of the maximum straining rate, $\left[\dfrac{d}{dt}(\Delta\varepsilon)\right]_{max}$, as a function of the stress σ, strain ε, and averaged strain rate $\dot\varepsilon_{avg}$. Referring to $(\dot{\overline{\Delta\varepsilon}})_{max}$ as the straining rate, and $\dot\varepsilon_{avg}$ as the strain rate, one notes experimentally

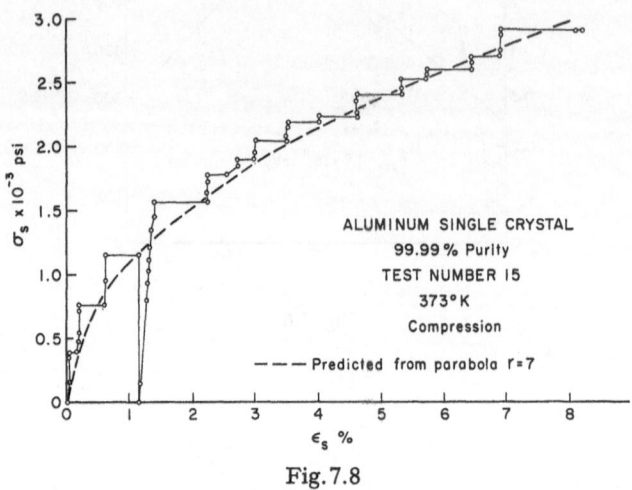

Fig. 7.8

that in a given experiment these two may differ by from 1 to 5 orders of magnitude. This is of particular importance for those who are attempting to determine a functional relationship between σ, ε, and $\dot\varepsilon$, since the measurements of $\dot\varepsilon$ in conventional machine testing is $\dot\varepsilon_{avg}$, whereas the actual plastic strain rate is $\dot{\overline{\Delta\varepsilon}}$.

In a constant stress rate experiment, the finite deformation is governed by a parabolic law such as Eq. (2.14). Differentiating with respect to time one finds that the strain rate, $\dot\varepsilon$, is proportional to the stress, Eq. (7.3):

$$\dot\varepsilon = \frac{2\dot\sigma}{\beta^2}\,\sigma. \tag{7.3}$$

Measurements of straining rate $(\dot{\overline{\Delta\varepsilon}})_{max}$ may be made directly from the slopes given on the chronograph tapes.

In Fig. 7.9 is shown (dashed line) the strain rate, $\dot\varepsilon_{avg}$, for the constant stress rate experiments of Figs. 3.3 and 3.4. These are compared with measured maximum straining rates, $(\dot{\overline{\Delta\varepsilon}})_{max}$, for the 120-hour dead-weight polycrystalline aluminum compression experiment of

Fig. 4.12 (circles), and for two of the constant tension stress rate experiments, 1219 and 1226.

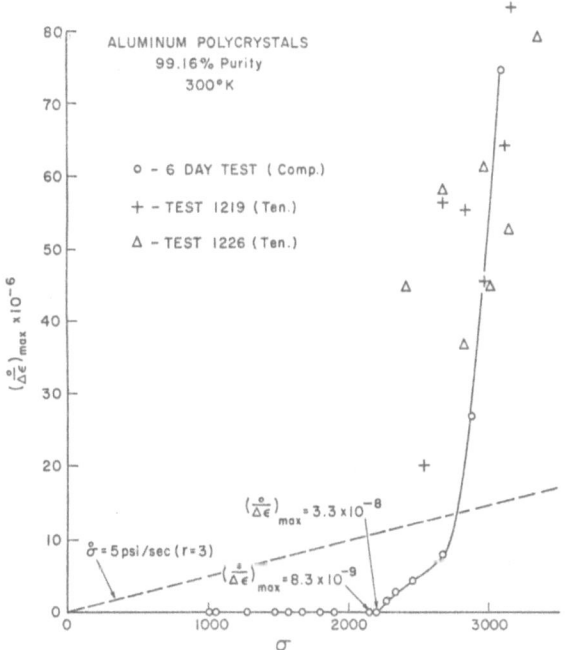

Fig. 7.9. The initiation of the Portevin-le Chatelier effect in an intermittently loaded compression experiment at $\dot{\varepsilon} = 10^{-8}$ sec^{-1} and in two constant stress rate tension experiments at strain rates of 10^{-5} sec^{-1} (dashed line). The ordinate is the maximum straining rate, $(\dot{\Delta\varepsilon})_{\mathbf{max}}$

For the 120-hour compression experiment Portevin - le Chatelier steps are observed at the elastic limit of $Y = 1,000$ psi and for all subsequent regions of strain. The detail of this 120-hour experiment in the very low strain region was described by BELL and STEIN (1962).

The point at which the maximum straining rate finally reached a value of 8.3×10^{-9} sec^{-1} is shown in Fig. 7.9. The maximum strain rates at lower stress levels fell far below this value. The next increment had a maximum straining rate of 3.3×10^{-8} sec^{-1}, with subsequent values increasing, as is seen in Fig. 7.9, up to the largest value at the maximum stress of 3,100 psi for this experiment (see Fig. 4.12).

The two constant tension experiments, 1219 and 1226, on the other hand, did not provide visible evidence of Portevin - le Chatelier steps until the average strain rate, $\dot{\varepsilon}_{\mathbf{avg}}$, intersected the solid line fared through the incremental loading 120-hour experiment. Once the Portevin - le

Chatelier effect had begun in these constant stress rate tension experiments, it continued to increase in approximately the same manner as
in the intermittent dead-weight compression experiment. In Fig. 7.10
is shown the continuation of these constant stress rate tension experiments, together with data from 5 other such measurements.

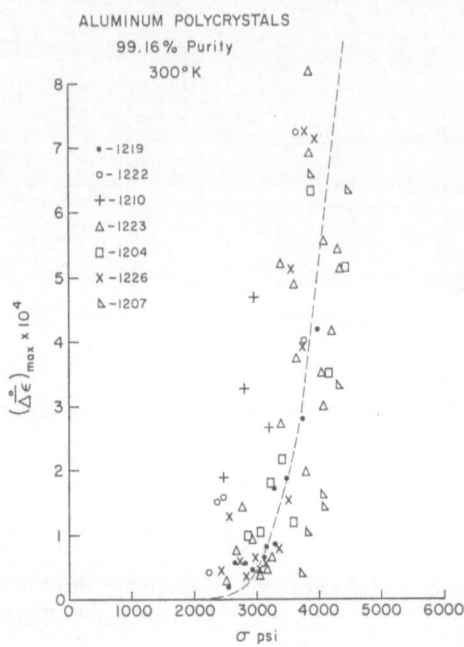

Fig. 7.10. Maximum straining rates vs stress for the initial Portevin - le
Chatelier steps in 7 constant stress rate tension experiments

The maximum straining rate of experiment 1219 is shown in Fig. 7.11,
which represents a continuation of the behavior shown in Fig. 7.10 to
higher stress and higher straining rates. One sees from these data that
the empirical approximation of Eq. (7.4) may be used to interrelate
maximum straining rate and stress.

$$(\bar{\Delta\dot{\varepsilon}})_{\max} = D\,(\sigma - \sigma_A) \tag{7.4}$$

where σ_A represents the intercept of the region of approximate linearity
upon the stress intercept. The value of D experimentally varies from
$8 \times 10^{-7}\ \sec^{-1}$ to $10.2 \times 10^{-7}\ \sec^{-1}$.

Introducing the aggregate theory ratios of Eq. (4.8) into Eq. (7.4)
provides the predicted maximum straining rate vs resolved stress
relation (dashed line) shown in Fig. 7.12.

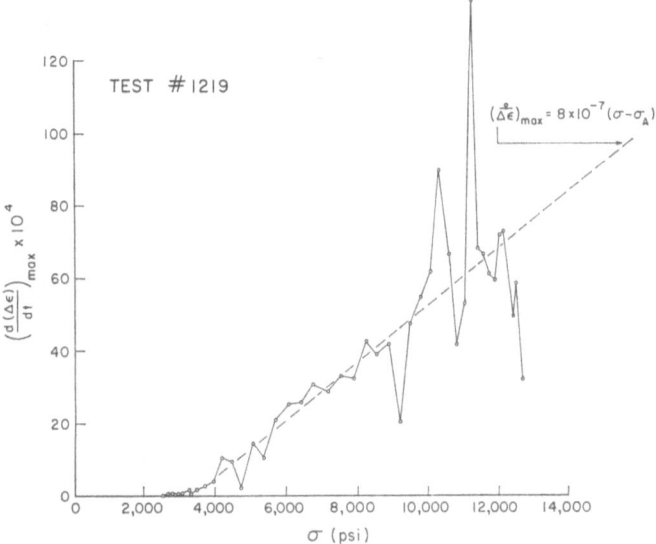

Fig. 7.11. Maximum straining rate vs stress for a constant stress rate tension experiment

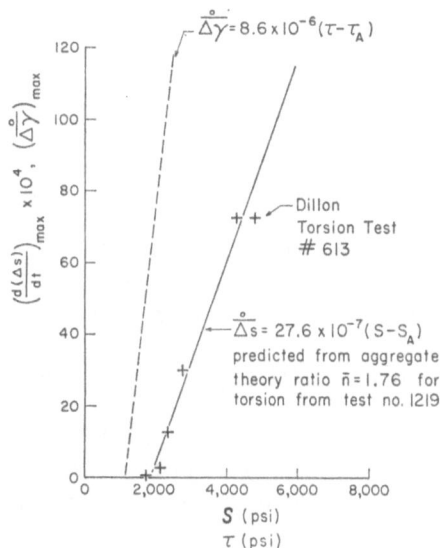

Fig. 7.12. Experimental maximum torsional shear straining rates vs torsional shear stress, compared with prediction from ratios of the aggregate theory, Eqs. (4.8) and (7.1)

The equation for the single crystal becomes:

$$(\overset{\cdot}{\Delta\gamma})_{\text{max}} = 8.6\times10^{-6}\,(\tau - \tau_A)\,. \tag{7.5}$$

Substituting this single crystal relation into the aggregate theory ratios for the torsion of hollow tubes, Eq.(7.1), one obtains the solid line shown in Fig.7.11, which may be compared with the experimental values of maximum straining rate, $(\overset{\cdot}{\Delta s})_{\text{max}}$, determined from the Δs-time data provided by DILLON (1963) for this experiment. Once again, the correlation for the aggregate theory ratio from the von Mises criterion of $\bar{n} = 1.76$ is operable.

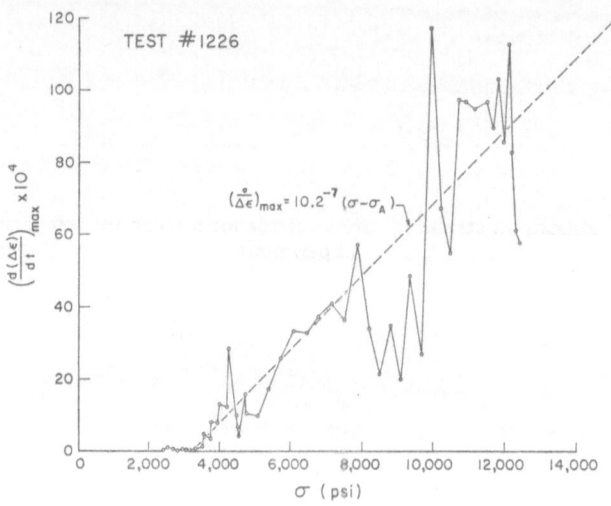

Fig. 7.13

One should also emphasize that here, too, in torsion the actual straining rates ranging from $1\times10^{-4}\,\text{sec}^{-1}$ to $7.5\times10^{-3}\,\text{sec}^{-1}$ are several orders of magnitude above the $\dot{\varepsilon}_{\text{avg}}$ of $2.6\times10^{-8}\,\text{sec}^{-1}$, for this Dillon test No.613.

As was shown in Chapter III, Fig.3.3, experiment No.1219 of Fig.7.12 is governed by a mode index $r = 3$ parabola to the second transition strain of 4.2%, occurring at a stress point 9,400 psi. In this experiment and in the identical experiment 1226 of Fig.7.13, also shown in Fig.3.3, a roughly reproducible behavior consistent with Eq.(7.5) is observed.

In experiment 1226 there is a nearly horizontal portion occurring at approximately 7,000 psi; at this point in experiment 1207 of Fig.7.14, a major change is observed. In experiment 1207 and in the identical experiment 1204 of Fig.7.15, the initial parabola has a mode index

$r = 2$ to the first transition strain of 1.5%, after which a transition occurs for each experiment to a mode index of $r = 3$ to the second transition strain of 4.2%. The stress corresponding to the first transition strain, $\varepsilon_N = 1.5\%$ is $\sigma_N = 6,800$ psi. The stress at the second transition strain of $\varepsilon_N = 4.2\%$ is $10,200$ psi. In both of these experiments of Figs. 7.14 and 7.15 one observes marked changes in the $(\overset{\bullet}{\Delta\varepsilon})_{\max}$ vs σ data at these stresses, corresponding to the first and second transition strains.

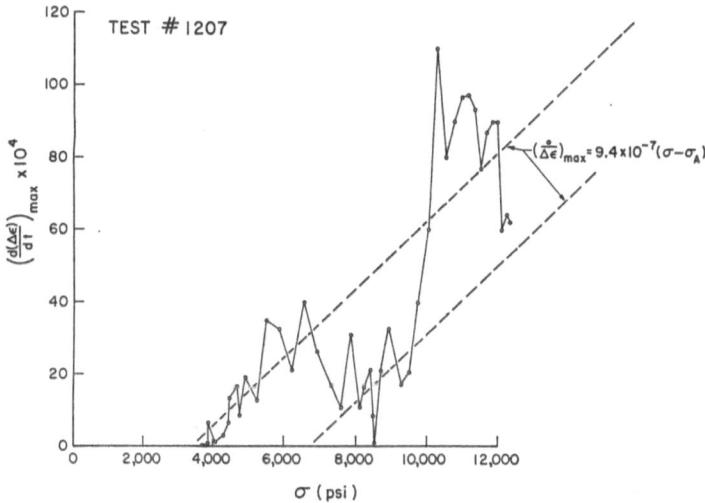

Fig. 7.14

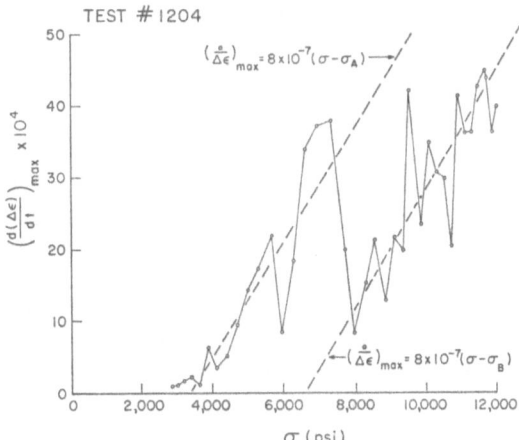

Fig. 7.15

For experiment 1204, the increasing slope in the second region is identical with that of experiment 1219 of Fig. 7.11, but the intercept, σ_B, now occurs at the stress for the first transition strain.

One experiment, 1210, shown in Fig. 7.16, has an initial $r = 3$ mode index distribution the same as that of experiment 1219 and 1226, but has a $(\dot{\overline{\Delta \varepsilon}})_{max}$ vs σ behavior more similar to that of experiments 1204 and 1207, with an initial mode index, $r = 2$.

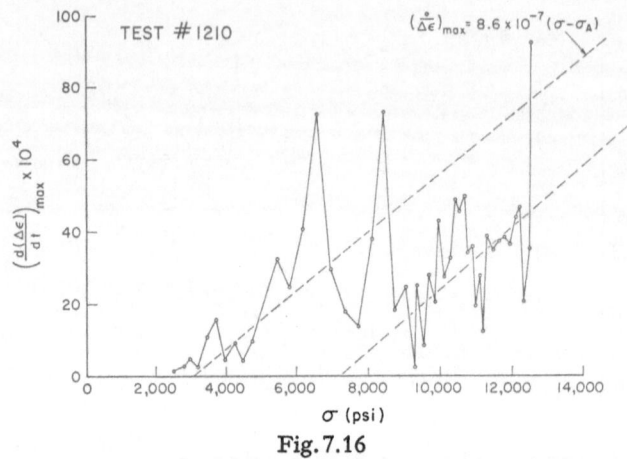

Fig. 7.16

SHARPE (1966 a, b, c) pointed out that several of his experiments exhibited a double-step behavior. He also showed that when these double steps were grouped, they might be compared with the more usual single step Portevin - le Chatelier behavior. Two experiments are shown in Figs. 7.17 and 7.18 for such double-step Portevin - le Chatelier behavior.

Although these data can be described as possessing double steps over certain limited ranges of the experiment, three distinct $(\dot{\overline{\Delta \varepsilon}})_{max}$ slopes were observable. These data are compared diagrammatically (dashed lines) with the experiments above, and with the average strain rate, $\dot{\varepsilon}_{avg}$, (dotted line). Both have initial $r = 3$ mode indices to the second transition strain of 4.2%, as may be seen in Fig. 3.3. The complexity of the straining rate phenomenon in this multiple step behavior demonstrates some of the difficulties which may be encountered in attempting to provide a theoretical basis for the Portevin - le Chatelier effect.

That this behavior is not confined to aluminum may be seen in Fig. 7.19 which shows the averaged Portevin - le Chatelier effect in three of HARTMAN's (1967) experiments in α-brass. These data not only exhibit the Portevin - le Chatelier effect superimposed upon a predicted parabola, but there is a $\Delta\varepsilon$ vs ε linear behavior similar to that in Eq.(7.2) for aluminum.

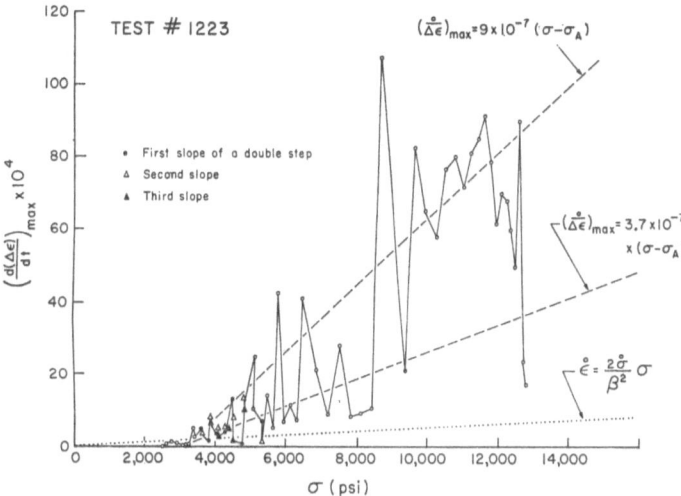

Fig. 7.17. Maximum **straining rate** vs stress for a double-stepped Portevin - le Chatelier effect experiment

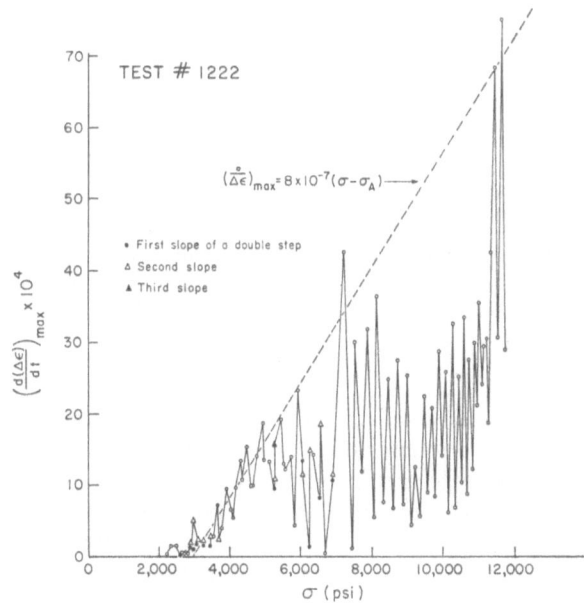

Fig. 7.18. Maximum straining rate vs stress for a double-stepped Portevin - le Chatelier effect experiment

As may be seen in Fig. 7.19 in the $\Delta \varepsilon$ vs ε plot for α-brass, the critical strains are precisely those obtained in aluminum.

Thus, the mode and transition structure of these two crystalline solids, one a pure metal and the other a binary combination of 70% copper and 30% zinc, demonstrate by the identity of their critical strains a quantitative and qualitative generality in the stability properties.

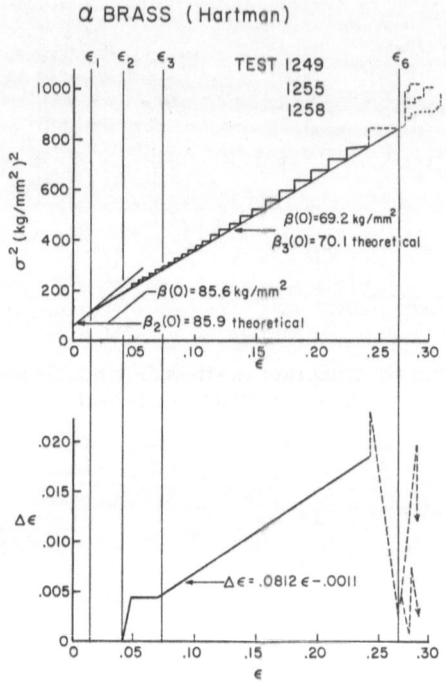

Fig. 7.19. Three averaged Portevin - le Chatelier experiments in α-brass, showing agreement with predicted parabolas, Eq. (3.1). Also shown are the changes occurring in the strain increment behavior at the critical strains of aluminum

The Portevin - le Chatelier effect occurs when the straining rate exceeds the average strain rate. This behavior is observable only when the average strain rate is sufficiently low so that the step behavior may be delineated experimentally. The observed phenomena are invariably superimposed upon the basic parabolic stress-strain function, and exhibit marked changes in stress increment $\Delta \sigma$, strain increment $\Delta \varepsilon$, and maximum strain rate $(\overline{\dot{\Delta \varepsilon}})_{max}$ at the critical, or transition, strains at which transitions in the mode indices have been observed.

One must emphasize again that the straining rate $\overline{\dot{\Delta\varepsilon}}$ is the actual strain rate of the plastic deformation, with most of the deformation time being spent in increasing along the vertical $\Delta\sigma$. The maximum straining rate, $(\overline{\dot{\Delta\varepsilon}})_{\mathrm{max}}$, may exceed the average strain rate, $\dot{\varepsilon}_{\mathrm{avg}}$, by several orders of magnitude. Therefore, one may understand some of the historical difficulties which have been encountered in attempting to construct stress-strain, strain-rate constitutive relations for metals.

The slope of the vertical step, $\Delta\sigma$, approximates E modulus values. Since the magnitude of this increment is very small (the maximum $\Delta\sigma$ for aluminum is 360 psi), it is difficult to make an accurate modulus determination from an individual slope. When a mode index change is taking place, such as from $r = 3$ to $r = 5$ at the second transition strain of 4.2%, it is possible to show that the averaged slopes of these vertical steps in the first region exceed those in the second region by the value of approximately 3/2, which would be consistent with the moduli which were proposed in Chapter VI.

A subtraction of $\Sigma\Delta\varepsilon$ from ε in a region in which a given mode index is known to be operable also has given numerical values consistent with the predicted moduli and moduli changes proposed in Chapter VI, thus providing further evidence that the mode index, r, and the moduli index, s, are different representations of the same phenomenon, and that finite distortional deformation mode transitions are accompanied by associated isotropic moduli transitons.

The $\Delta\sigma$ vs ε behavior of Fig. 7.2 may be varied by modifying experimental conditions. For α-brass, $\Delta\sigma$ is nearly a constant function of ε. R. MITTAL, in current experiments in the present writer's laboratory, has been able to produce experimental modifications of the straining rate behavior for polycrystalline aluminum. The Portevin - le Chatelier effect, whatever its variation of detail, always is superimposed upon the basic parabolic stress-strain function of the present generalization which applies whether this phenomenon is present or not. The phenomenon, even though undergoing major modification at transition strains, must, therefore, play a secondary role as far as present experimental evidence would indicate.

Chapter VIII

Binary Combinations

The finite distortional deformation of the elements described in the previous chapters is specified completely for a given test temperature when the stable zero-point isotropic shear modulus, the melting point, and the mode index, r, are known. Nowhere in this formulation does crystal structure appear to play a significant role in finite distortion. An obvious next step is to consider the finite distortional behavior of mixtures of elements to determine whether or not additional factors must be introduced as a function of the percentage composition of the components.

Choosing a binary alloy each of whose components has widely different zero-point isotropic shear moduli and/or melting points offers a particularly interesting situation. When the percentage composition is varied from one pure component to the other, changes in the distribution integers, s and p, must occur if the generalization is still to apply. The opportunity thus is offered for observing some of the factors which give rise to the discrete distribution.

Binary alloys such as α-brass are thought to have a dislocation stacking fault energy nearly an order of magnitude lower than, say, that of aluminum. From the point of view of dislocation theory, if stage I and stage II behavior concepts are extrapolated to stage III, this difference in stacking fault energy should have a marked effect upon the finite distortion behavior. To demonstrate, as HARTMAN (1967) has done in the binary combination 70—30 α-brass, that the present parabolic finite distortional deformation generalization is applicable in every respect, has obvious theoretical significance.

A series of uniaxial stress tension experiments of SIMMONS (1961) in yttrium polycrystals at 300° K are shown in Fig. 8.1.

In Chapter III unalloyed yttrium was shown to have a parabolic mode and deformation structure consistent with the present generalization. SIMMONS' data in Fig. 8.1 was "de-trued" for a comparison with the present writer's parabolic generalization prediction for the originally measured *nominal* stress-strain functions. In order to compare these data with each other and with the earlier quasi-static data of

Chapter III, they have been projected to absolute zero. Hence, Fig. 8.1 is plotted as:

$$\left(\frac{\sigma}{1 - T/T_m}\right)^2 \text{ vs } \varepsilon.$$

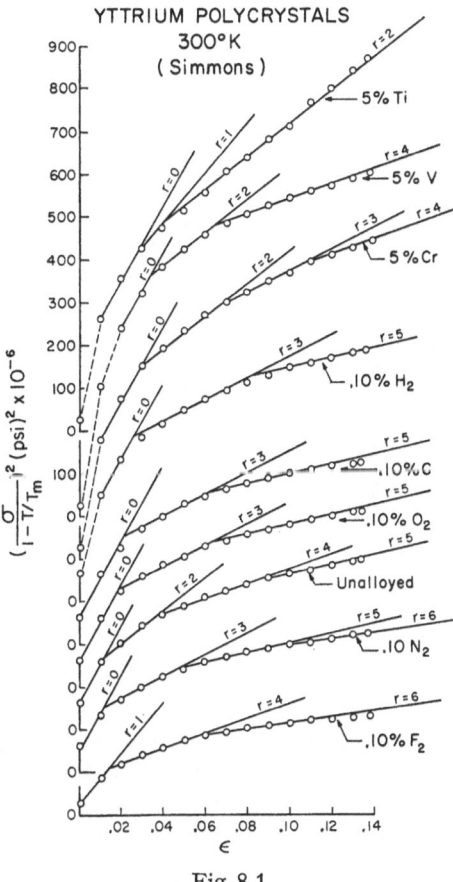

Fig. 8.1

Despite the variety of impurities and the addition of 5% Ti or V, or Cr, it may be seen from Fig. 8.1 that yttrium binary combinations are closely given by the present parabolic large deformation generalization. It is interesting that the initial parabolic large deformation mode index is $r = 0$ in all but one instance. One wonders whether the stable zero-point isotropic shear modulus for this element is $s = 11$, $p = 1$ as given in Chapter V, or whether this modulus measurement is one of the multiple elastic values of Chapter VI. The first transition from the initial parabola for the five lower experiments in Fig. 8.1 all occur in the vicinity of the first aluminum transition strain of 1.5%.

HARTMAN's α-brass data is for the polycrystalline solid. As was shown in Chapter II, his experimental diffraction grating studies of this crystalline solid reveal that its finite amplitude wave propagation

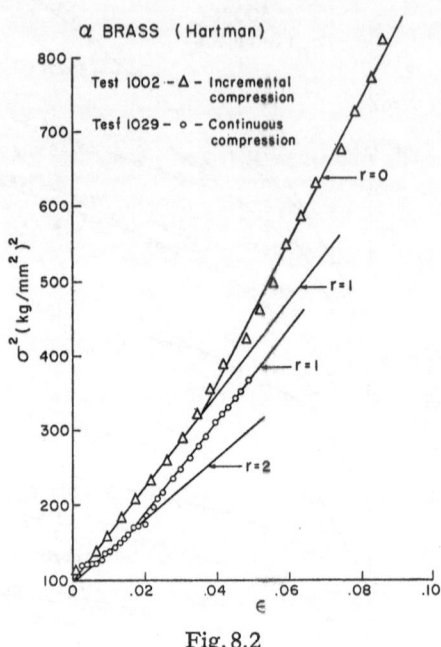

Fig. 8.2

characteristics are in very close agreement with the present generalization. The multiple linear elasticities occur in the same form for this binary combination as for the single elements considered in Chapter VI; furthermore, for α-brass the Portevin - le Chatelier phenomenon not only is present but is superimposed upon a base parabola of the present generalization, as was shown in Chapter VII. Now, by presenting typical examples, Fig. 8.2, of quasi-static uniaxial compression measurements one may see that for this situation, the parabolic generalization is applicable. (Also see Fig. 7.19.)

The predominant parabolic large deformation mode index for 70—30 α-brass is $r = 3$ both for dynamic and for quasi-static data (see Fig. 7.19). However, a wider variety of initial mode indices (from $r = 1$ to $r = 5$) is observed experimentally in the dynamic deformation of this annealed solid than is found in either the element copper or the element zinc of which it is composed. This distribution of diffraction grating measured parabola coefficients for 24 symmetrical free-flight impact experiments of HARTMAN (1967) is shown in Fig. 8.3. The agreement of the averaged data with prediction is indeed remarkable.

70 - 30 α-Brass (Hartman)

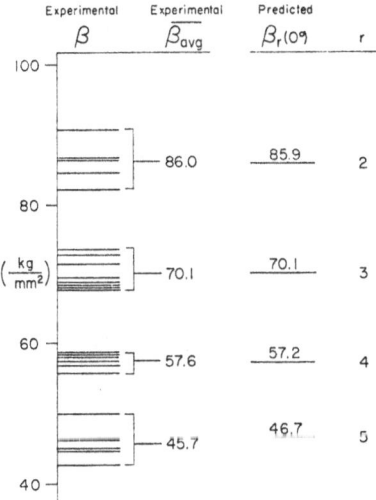

Fig. 8.3. Experimental parabola coefficients from diffraction grating wave speed determinations. The averaged values of each group are compared with prediction from Eq. (2.20) for the mode index, r, shown

HARTMAN (1967) has discussed some of the prior deformation factors which distinguish one mode index from another in otherwise identical impact experiments. He has suggested that the difference is due to α-brass having a sensitivity greater than that of either of these two elements or of aluminum, α-brass being more sensitive to the prior thermal or mechanical history. HARTMAN has made a detailed study of the single crystal brass stage III deformation data of THORNTON and MITCHELL (1963) for various percentage compositions including 70—30 α-brass. He has shown that in these data the distribution of $\beta_{ro}(II)$ parabola coefficients is consistent with prediction from the introduction of α-brass polycrystal parabola coefficients into the ratios of the aggregate theory, Eq. (4.8). He also has shown that stage II and stage III deformation of α-brass single crystals are related in terms of Eq. (4.5).

An interesting series of single crystal stage III deformation experiments in Au—Ag for various percentage compositions was made over 35 years ago by SACHS and WEERTS (1930). Silver and gold have nearly identical zero-point isotropic shear moduli (see Chapter V) and also quite similar melting points, 1,234° K and 1,336° K, respectively. Therefore, the parabola coefficients at a single temperature such as room temperature, unless differing by a change in the mode index r as a function of composition, would be expected to be not too different. The well-known variation in the elastic limit or critical resolved shear stress of a binary alloy as a function of composition would not be expected to be of importance in the present finite deformation behavior.

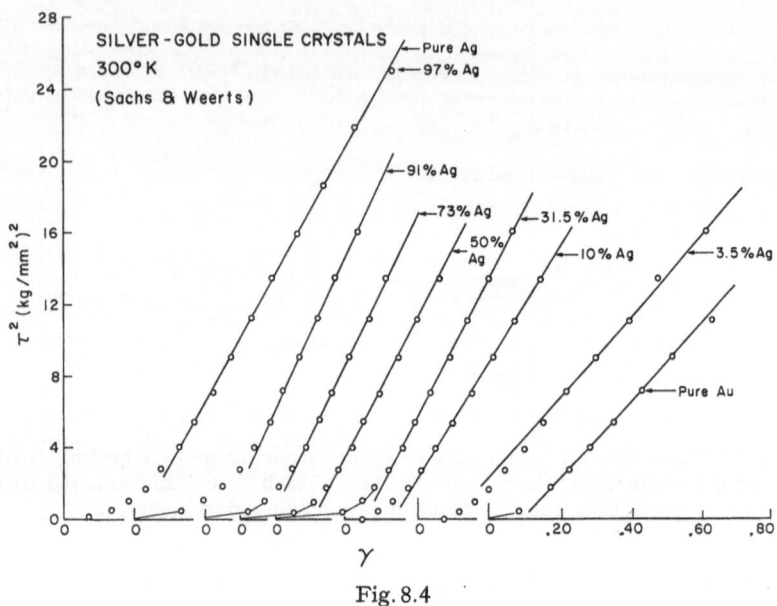

Fig. 8.4

In Fig. 8.4 are shown the τ^2 vs γ plots of these data of Sachs and Weerts for the designated percentage combinations. Stage III parabola coefficients $\beta_{ro}(\text{II})$ calculated from introducing stage II parameters in Eq. (4.3), and directly measured stage III zero-point parabola coefficients $\beta_{ro}(\text{III})$, are shown in Table XII from which it may be seen that silver and the binary combination Au—Ag, irrespective of percentage composition, furnished finite deformation stress-strain functions consistent with the parabolic large deformation generalization.

The binary combination Ni-Co, like that of Au-Ag, has a very closely matching zero-point isotropic shear modulus and melting point. Therefore, one would expect that in this solid there would be very small variation of parabola coefficients with percentage composition. The experimental data considered are those of Kronmüller (1959) in nickel and Ni-Co 20 and of MEISSNER (1959) in nickel and Ni-Co with 10, 20, 30, and 40% cobalt. The τ^2 vs γ plots of these data of KRONMÜLLER and MEISSNER are shown in Fig. 8.5.

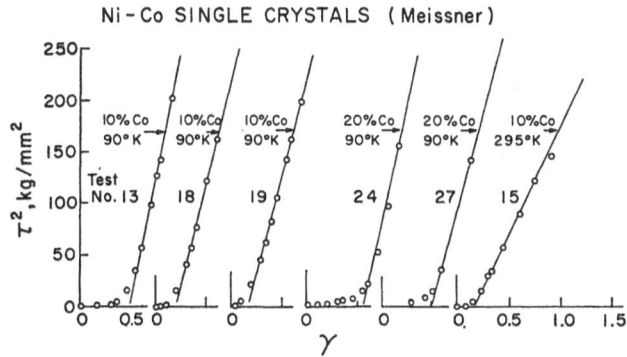

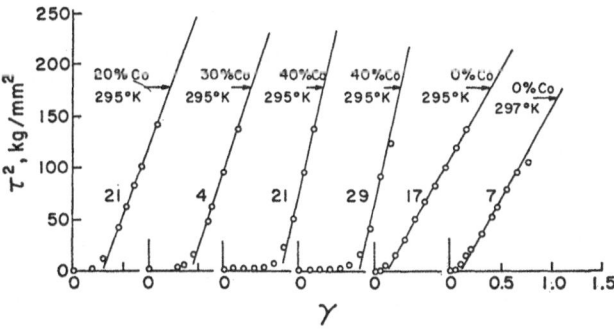

Fig. 8.5

Table XII. *Ag-Au (Sachs and Weerts) 300° K*

% Comp. Silver	τ^*	θ_{II}	$\beta_{ro}(II)$	γ	$\beta_{ro}(III)$	γ
0	—	—	—	—	5.97	5
3.5	—	—	—	—	6.07	5
10	3.9	9.09	6.76	4	7.35	4
31.5	1.55	11.94	7.85	4	8.16	3
50	2.38	10.75	9.25	3	8.17	3
73	2.8	10.0	9.73	3	8.23	3
91	3.0	9.09	9.66	3	8.48	3
97	2.65	8.55	8.89	3	7.89	4
100	—	—	—	—	7.90	4

Calculated $\beta_{ro}(II)$ and $\beta_{ro}(III)$ stage III parabola coefficients for these data are listed in Table XIII. These experiments were performed at 90° K and 295° K.

Table XIII. *Ni-Co*

°K Temp.	% Comp. Cobalt	Test	τ^*	θ_{II}	$\beta_{ro}(II)$	r	$\beta_{ro}(III)$	r	Reference
90	0	18	9.2	23.9	22.13	4	21.32	4	Meissner
90	0	19	9.5	22.5	21.81	4	20.66	4	Meissner
295	0	17	1.56	22.3	10.04	7	15.95	5	Meissner
297	0	7	2.32	23.5	12.59	6	16.21	5	Meissner
90	10	13	11.0	23.6	24.03	3	16.68	5	Meissner
295	10	15	2.6	23.6	13.35	6	17.47	5	Meissner
90	20	24	12.6	24.9	26.43	3	22.38	3	Meissner
90	20	27	12.0	22.4	24.45	3	20.76	4	Meissner
90	20	—	11.5	27.5	20.05	4	19.90	—	Kronmüller
295	20	—	4.0	22.3	19.40	4	18.90	—	Kronmüller
295	20	21	4.04	21.5	15.88	5	19.87	4	Meissner
295	20	25	4.2	23.2	16.8	5	—	—	Meissner
90	30	7	12.0	27.7	27.2	2	—	—	Meissner
295	30	4	5.94	21.3	19.18	4	21.12	4	Meissner
90	40	30	4.55	22.7	23.0	3	—	—	Meissner
90	40	15	4.0	26.0	23.4	3	—	—	Meissner
295	40	21	11.25	24.0	28.02	2	25.11	3	Meissner
295	40	29	10.0	23.6	26.20	3	25.11	3	Meissner

As would be expected from the data considered in Chapters III and IV, the mode index r decreases with decreasing temperature; i.e., the parabola coefficient is higher at lower temperatures. What also may be seen in these Ni-Co data is that the parabola coefficient increases with an increasing percentage of cobalt. The general agreement of the deformation behavior of this binary combination with the parabolic generalization suggests that the study of initial mode indices as a function of percentage composition probably will be a fruitful undertaking.

All of the solids thus far discussed in this monograph have been metals. The writer is greatly interested in determining whether or not the generalization is limited to that group of crystalline solids. At the present time insufficient large deformation data in non-metallic single crystals and polycrystals is available, so it is not yet possible to determine what might be expected in such situations. One exception to this sparsity of data is the parabolic data of Thiele (1932) in NaCl. Taylor in his 1938 study of parabolicity in the plasticity of single crystals calculated parabola coefficients for Thiele's single crystal data. Unfortunately, Thiele's data cannot be resolved since no x-ray diffraction measurements of initial orientations were provided. The σ_s^2 vs ε_s plots of the single crystal data are shown in Fig. 8.6. By the straightness of the data the expected parabolicity is demonstrated.

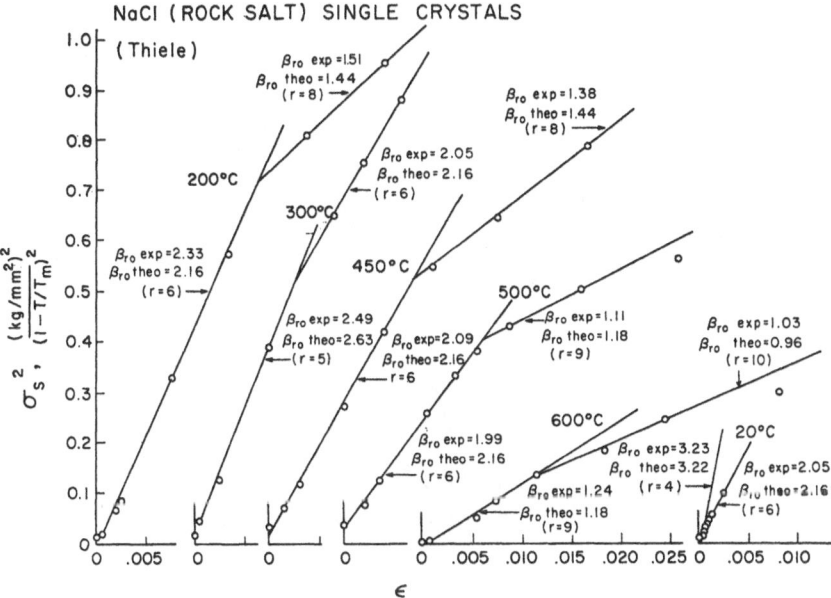

Fig. 8.6. Parabolic deformation in rocksalt single crystals, compared with prediction, Eq. (4.11), assuming central crystallographic orientations. Note that the ratios of initial and transition experimental parabola coefficients occur in the predicted $(2/3)^{r/2}$ distribution

The measured parabola coefficients for each temperature have been used to calculate zero-point parabola coefficients using a melting point for NaCl of $T_m = 1,074°$ K. From the zero-point shear modulus of Chapter V, $\mu(0) = 1,381$ kg/mm² ($s = 15$, $p = 0$) and assuming for the sake of comparison an initial central orientation in each instance, predicted zero-point parabola coefficients for the designated mode indices r are given in Fig. 8.6. Even though THIELE did not provide crystallographic data so that a close correlation with prediction could be established, the observed finite distortion of NaCl is sufficiently good to lend weight to the conjecture that non-metallic crystalline solids may be expected to behave in finite distortion in a manner not too different from metallic crystals. There is sufficient agreement in these rocksalt data to indicate the desirability of a major research program in non-metallic crystals in relation to the present generalization.

To effectively study changes in the distribution integers, s and p, by varying the percentage composition of a binary alloy for which both pure elements are expressible in terms of the present parabolic finite deformation generalization, it is desirable that the zero-point isotropic shear moduli of the components should be widely different. The only binary combination considered in this chapter meeting this condition is molybdenum-50% rhenium.

For molybdenum $\mu(0) = 15,400 \text{ kg/mm}^2$ ($s = 3$, $p = 0$), and for rhenium $\mu(0) = 21,400 \text{ kg/mm}^2$ ($s = 1$, $p = 1$). The melting points are

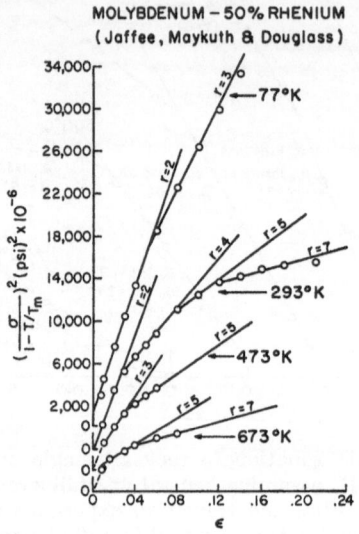

Fig. 8.7. Parabolic finite deformation in molybdenum-50% rhenium poly-crystals (circles), compared with prediction, Eq. (3.1), using $\mu(0)$ and T_m of rhenium (solid lines)

high for both elements, being 2,893° K and 3,442° K, respectively. Both of the pure components of this binary combination were shown in Chapter III to have finite distortional deformation in close agreement with the present parabolic generalization stress-strain function (see Figs. 3.13, 3.21, 3.22, and 3.23).

JAFFEE, MAYKUTH, and DOUGLASS (1961) have provided 4 poly-crystalline uniaxial stress experiments in Mo-50 Re, at temperatures ranging from 77° K to 673° K. These data are shown in Fig. 8.7 in a $\sigma^2/(1 - T/T_m)^2$ vs ε plot which had to be "de-trued" for comparison with prediction. It may be observed that predicted parabolicity and mode and transition structure and linear temperature dependence are present.

One does not know in advance which mode indices are applicable since four possibilities exist from rhenium to molybdenum; i.e., $s = 1$, $p = 1$; $s = 2$, $p = 0$; $s = 2$, $p = 1$; $s = 3$, $p = 0$. Predicted values of the zero-point parabola coefficients for each of these four groupings for values of the mode index, r, from $r = 1$ to $r = 7$, are shown in Table XIV, where the 5 experimentally observed parabolas of Fig. 8.7 are in italics.

Table XIV. *Mo-50 Re polycrystalline parabola coefficient* $\beta(0)$

Mode Index r	$s = 1, p = 1$ kg/mm²	$s = 2, p = 0$ kg/mm²	$s = 2, p = 1$ kg/mm²	$s = 3, p = 0$ kg/mm²
1	486	440	*397*	360
2	*397*	360	*325*	294
3	*325*	294	*265*	240
4	*265*	240	*217*	196
5	*217*	196	177	161
6	177	161	*145*	131
7	*145*	131	118	107

On examining the table, there is no question that combining these two elements in equal proportions provides a value of $p = 1$ in agreement with that of rhenium and not with that of molybdenum. One cannot be equally certain as to whether the integer, s, is $s = 1$, or $s = 2$. The mode indices shown in Fig. 8.7 are those for $s = 1$, $p = 1$ of the pure component, rhenium. The highest experimental parabolas with the deformation mode index $r = 2$ at 77° K and 293° K exceed the maximum value for molybdenum for $s = 3$, $p = 0$, and $r = 1$.

If one designates the discrete distribution of the finite deformation parabola coefficients by changes in the integer, s, rather than by the introduction of the additional mode index integer, r, as was proposed in Chapter VI, then these parabola coefficients of Fig. 8.7 and Table XIV would be expressible in terms of a series of increasing integers beginning with $s = 3$, for which $p = 1$ in every instance.

Rhenium has a hexagonal crystal structure and molybdenum a body-centered cubic crystalline structure. The value of p in these Mo-50 Re data is given by the hexagonal component. The Cu-30 Zn α-brass of HARTMAN (1967) from which the face-centered cubic component copper has indices $s = 8$, $p = 1$, and hexagonal zinc, $s = 9$, $p = 0$, also provides parabola coefficients and a zero-point isotropic linear elastic shear modulus agreeing with the hexagonal zinc component.

Both silver and gold have the same integral indices, $s = 11$, $p = 0$, and consistent parabola coefficients, as was shown above in this chapter.

Both of these elements are face-centered cubic. The nickel-cobalt single
crystals, also described above, have pure component moduli which
differ in the integral index, p. For nickel, $s = 6$, $p = 0$, and for cobalt
$s = 6$, $p = 1$.

All of the mode indices for the nickel-cobalt single crystal data in
Table XIII, with one possible exception, provided parabola coefficients
agreeing with the face-centered cubic nickel. (The values of r listed in
Table XIII were determined from the nickel theoretical parabola
coefficients.) HARTMAN (1967) in analyzing the brass single crystal
experiments of MITCHELL and THORNTON (1963) which varied from pure
copper to 30% zinc, showed that up to 10% zinc the governing integral
indices were $s = 8$, $p = 1$, as in pure copper. From 15% to 30% zinc
composition, the integral indices were $s = 9$, $p = 0$; i.e., a change of
p from $p = 1$ to $p = 0$ occurred as the percentage composition was
changed.

Independent of the particular indices which govern the finite deformation of a particular composition, all of these data, whether single
crystal or polycrystal, were experimentally found to be parabolic with
linear temperature dependent parabola coefficients and to fall within
the framework of the deformation mode and transition structure of
interest in this monograph. It seems likely that a further study of the
parabola coefficients, as well as the mode and transition structure as a
function of percentage composition in binary alloys, will provide a
fruitful approach to an understanding of the discretely distributed
states in which finite distortional deformation is found to occur.

The experiments described in this chapter by no means include all
the binary combination data available in the literature. [See, for example,
the experimental studies of PHILLIPS (1953), PHILLIPS, SWAIN, and
EBORALL (1953), and KRUPNIK and FORD (1953) on the Portevin-
le Chatelier effect in various alloys of aluminum; or the numerous
standard studies and surveys on the finite deformation of metal alloys.]
The experiments here, particularly those of HARTMAN (1967) in α-brass,
are sufficient to demonstrate that every aspect of the present large
deformation parabolic generalization and transition structure does apply
to binary combinations of elements, independent of percentage composition, and independent of differences in the stacking fault energy, elastic limit, melting point, and specific crystal structure.

Chapter IX

Transition Phenomena

In 1960 when the writer's experimental finite amplitude wave propagation studies were successfully extended to higher impact velocities it became apparent that the governing stress-strain function at higher strains in general differed from quasi-static stress-strain functions determined at strain rates 7 or 8 orders of magnitude lower. This difference above the first transition strain, now known to be derivative from the experimental fact that parabola transitions do not occur during the rise of a finite amplitude wave front, was the genesis of the study which led to the discovery of the mode and transition structure of the present parabolic generalization. The problem of studying such behavior was complicated by the fact that for high purity annealed aluminum the dynamic and quasi-static stress-strain functions were the same to very large strains; i.e., quasi-static transitions first occurred at large transition strain.

It was early observed (BELL, 1961b) that the ratio of high and low purity dynamic aluminum experimental parabola coefficients was numerically equal to an empirical factor $(2/3)^{3/2}$. A similar behavior was observed in comparing the resolved stage III deformation of cubic single crystals of varying purity (BELL, 1961b; BELL and WERNER, 1962). In addition, some solids such as low purity zinc had a higher quasi-static than dynamic stress-strain function, while for solids such as magnesium with an upward turning quasi-static function, no correlation at all seemed to exist. At the same time that these parabolic stress-strain functions were first being studied, the series of transition velocities described in Chapter II were discovered (BELL, 1961b, 1962a, 1963b; BELL and SUCKLING, 1962). Between 1962 and the present the writer gradually realized that all of these phenomena were interrelated in the form of the discrete deformation mode and transition structure which has been described in the previous 8 chapters of this monograph. Many of the experiments described above were performed for the purpose of determining whether or not the transition parameter was stress, particle velocity, or strain. Experimental evidence that the transition behavior occurred at fixed values of each of these variables was accumulated. The discovery that the transition strains were independent of the mode indices of the two parabola coefficients on either side of the

transition eliminated stress and particle velocity as the governing parameters. The additional discovery that polycrystalline transition strains
had their counterpart in the resolved stage III deformation of the
single crystal through aggregate theory ratio further eliminated the
particle velocity as the transition parameter.

Single crystal mode indices for the 8 elements and 4 binary combinations considered in this monograph vary from $r = 1$ to $r = 12$ with
the preponderance of the data falling in the range of $r = 4$ to $r = 8$.
A comparison of numerical distributions of the 602 polycrystal parabola
coefficients and 781 single crystal parabola coefficients of earlier chapters
is shown in Fig. 9.1.

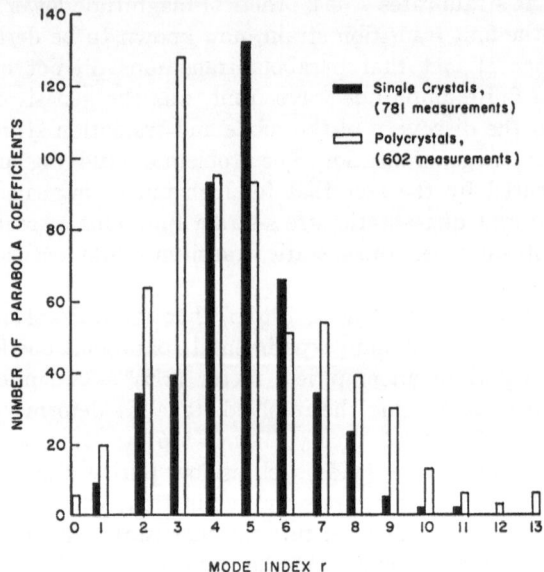

Fig. 9.1. The distribution of experimental single crystal and polycrystal
initial and transition parabola coefficients among the mode indices, r. Note
that differences in fractional melting point distributions and purity distributions have not been delineated

The polycrystalline data for the 19 elements and numerous binary
combinations considered have parabola coefficients which vary from
$r = 0$ to $r = 13$, with the preponderance of the data falling between
$r = 2$ and $r = 8$. In constructing the mode index description of these
experimental data, no value of r was introduced until it had been
established both in polycrystalline experiments and in single crystal
experiments. The magnitude of the variation in parabola coefficients
between one mode index and another is more than sufficient to demon-

strate that the stress-strain functions of these experiments, both individually and collectively, are very closely given in Eqs. (3.1) and (4.11) for the polycrystal and single crystal finite distortional deformation, respectively.

In considering the experimental data of the first 8 chapters, one may show that the numerical value of the initial mode index r depends upon the ambient temperature, purity, strain rate, elastic working prior to the inauguration of large deformation, predeformation temperature history (annealing temperatures, etc.) and the prior large deformation history. It should be made emphatically clear that although in a few instances the value of the initial mode index r may be set by the rate of strain imposed in uniaxial stress tests for which the deformation history is prescribed, there is no observable dependence of the governing stress-strain function upon viscosity during the large deformation.

When the finite distortion for an initial parabola has reached a transition strain, one of three possibilities is present: 1. the deformation may continue to be governed by the same mode index; 2. the mode index may increase; and 3. the mode index may decrease. In the second and third situations the Δr of the second-order transition may involve a change of one, two, or more integral values. The value of Δr in such transitions depends upon the transition strain ε_N or γ_N under consideration, the purity, the ambient temperature of the experiment, the type of test (tension or compression), and the strain rate. The range of strain in the average finite deformation mode is relatively large as is the number of experiments in which no transition occurs. An inspection of the data of Chapters III, IV, and VIII reveals that the number of experiments with more than two transitions is relatively small.

The manner in which the average strain rate $\dot{\varepsilon}_{avg}$ is related to the transition structure is shown in Fig. 9.2, based upon the uniaxial stress experiments in Chapters II and III.

In the dynamic data of Chapter II, from diffraction grating measurements and hundreds of wave propagation experiments, the deformation mode index has been shown to proceed through each of the first three critical strains without a transition or change in the deformation mode index. The quasi-static stress-strain function of the uniaxial compression experiments, on the other hand, do exhibit transitions as the deformation increases, as for example the experiments (Fig. 3.4) designated as "Avg 1957, 1, 2, 3" whose behavior is shown in the lowest curve of Fig. 9.2. The transitions occur at the designated transition strains ε_N from the initial value of $r = 2$, which is the mode index of the entire dynamic deformation, to successively, $r = 3$, $r = 4$, and $r = 5$.

Also shown schematically (dashed lines) in Fig. 9.2 are situations in which no transition takes place at the first critical strain, but mode indices changes do take place at the second and third critical strains. Also shown is the situation in which the first transition strain occurs at the third critical strain. Numerous examples of behavior of Fig. 9.2 are shown in the nearly 300 polycrystalline quasi-static uniaxial stress experiments in Chapters III and VIII, in 19 elements and several binary combinations. Several of the high purity aluminum polycrystals of Fig. 3.5 deform with parabolas of initial mode indices $r = 5$ or $r = 6$

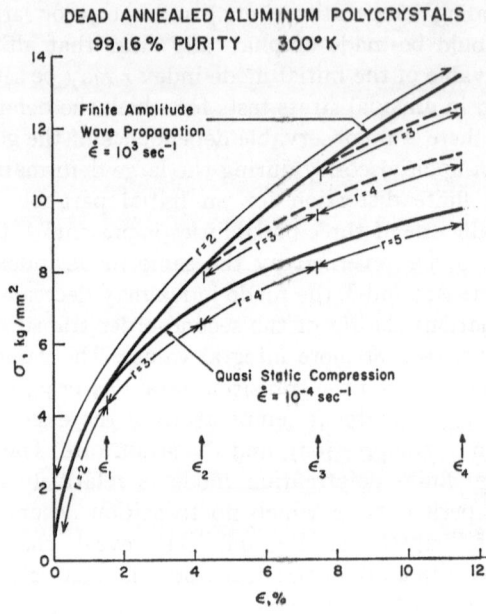

Fig. 9.2. The effect of strain rate upon finite deformation mode stability at critical strains

continuing to the third or fourth critical strain before a transition occurs. It is important to emphasize that the critical strains in general are independent of the entering mode indices.

In view of the numerous combinations of deformation modes which may occur and of the fact that whether or not a particular transition occurs at a critical strain is related to the initial deformation mode index, it is not too surprising that proposed definitions of viscosity have seemed experimentally plausible when they were based upon a variation in the magnitude of a measured stress at a measured strain at different strain rates. The over 900 uniaxial stress experiments for both polycrystals and cubic single crystals, described in the first eight chapters of this monograph, experimentally establish that such a variation in the

magnitude of the stress, when it occurs, arises from the stability properties of a discretely distributed finite deformation mode and transition
structure rather than from viscosity. The constitutive relation does *not*
depend upon the strain rate $\dot{\varepsilon}$.

A large number of experiments were described in Chapter III in
aluminum and α-brass which were carried out for the purpose of studying
this transition phenomenon. Varying such parameters as prior thermal
history, prior deformation, etc., alters both the magnitude of the initial
mode index and subsequent changes in the mode index. The experimental
study of this phenomenon reported in this monograph for the first time,
is still very much in the beginning phases. Any consideration of this
behavior must take cognizance of the experimental fact, shown in
Chapter VI, that there are similar and obviously related transitions in
elastic moduli whose zero-point values have the same distribution as
that of zero-point parabola coefficients.

That the polycrystalline transition strains are the same in the
element aluminum and in the binary alloy α-brass suggests a degree of
quantitative generality which at this time can only be stated as an
experimental fact. Polycrystalline critical strains of 1.5, 4.2, 7.5, 11.5,
16.3, and 26.6% and their cubic single crystalline counterparts of 4.6,
12.9, 23, 35, 50.5, and 81.4% are experimentally related through the
ratios of the Taylor (1938) aggregate theory, Eq. (4.8), as has been shown
in a very large number of examples in Chapters III and IV.

The dynamic wave propagation critical velocities which were discovered before the existence of their quasi-static counterparts was
realized, have for the first three values: 1,478 cm/sec, 3,350 cm/sec, and
5,080 cm/sec, corresponding, through Eq. (2.21), to the first three
critical strains for the dynamic mode index of $r = 2$ in polycrystalline
aluminum.

One notes empirically that the ratio of the second critical velocity
to the first is 2.26, and that of the third critical velocity to the second
is 1.53 i.e., $(3/2)^2$ and $(3/2)$, respectively. These values of velocity are for
a fixed mode index of $r = 2$, but that they occur in this form suggests
that the critical strains themselves, or the difference between one $(\varepsilon_N)^{3/4}$
and another, empirically falls into some sequence related to the finite
deformation mode and transition structure itself.

Most theoretical treatments of quasi-static plasticity in metals are
very much concerned with describing yield criteria or the discontinuity
between the small linear elastic initial region and the region of permanent
deformation at the elastic limit, Y. A very large fraction of the mathematical plasticity literature has been concerned with the generalization
of such yield criteria in terms of small stress and strain increments in
the immediate vicinity of yield surfaces.

When prior work hardening, prior thermal history, or the variation of the percentage composition of binary alloys produce changes in this elastic limit, Y, the opportunity is offered to examine the effect of such variations upon the present generalization. An inspection of the several hundred polycrystalline experiments in Chapters III, VI, and VIII reveals that the observed data are not described by an equality in Eq.(9.1) as is sometimes proposed. In fact one must write:

$$(\sigma - Y)^3 \neq \beta^2 \varepsilon. \qquad (9.1)$$

A large number of experiments are given in Chapters II to VIII in which such changes of the elastic limit stress, Y, appear. As illustrative examples one might refer to the gold data of Fig.3.14, or the copper data of Fig.3.17, in which completely annealed experiments are compared with those for which there has been a prior deformation of 5%.

All of these many experiments in which the elastic limit Y varies reveal that the essential effect of changing the elastic limit is to shift the intercept ε_b on the strain abscissa, Eq.(9.2).

$$\sigma^2 = \beta^2 (\varepsilon - \varepsilon_b). \qquad (9.2)$$

This behavior of Eq.(9.2) shown schematically in Fig.9.3a *does* apply to all of these data.

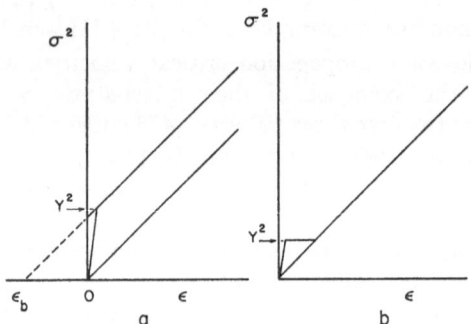

Fig.9.3. (a) The observed comparative parabolic deformation behavior of completely annealed and preworked crystalline solids. (b) The experimentally observed initial behavior of a completely annealed solid with an elastic limit, Y

In this connection, it is very important to note that dynamic wave propagation studies of Al, Zn, Mg, and α-brass in which experiments have been compared with major differences in the elastic limit, Y, the final plastic strain determined from the finite amplitude wave theory is unaffected by the variation in the elastic limit, Y. In both dynamic and quasi-static *completely annealed* crystalline solids the presence of an elastic limit not produced by prior workhardening is as is shown in Fig. 9.3b; i.e., the finite distortion is not affected by the magnitude of Y.

As has been indicated many times above, the generalized parabolic stress-strain function described in this monograph is for the nominal or engineering stress-strain function and refers to the undeformed state of the solid. An enormous number of "de-truing" calculations had to be made in order to compare the polycrystalline data from the literature, with prediction. In this writer's opinion, the existence of this generalization was obscured by the common practice of recalculating the measured uniaxial stress-strain data as reduced stress and logarithmic strain. When one attempts to compare stress-strain functions with slightly different mode and transition structures in terms of logarithmic strain, it becomes readily apparent why a finite distortional generalization for crystalline solids was not detectable.

As was shown in Chapter IV, through the ratios of the aggregate theory both the polycrystalline uniaxial stress-strain and the single crystal uniaxial stress-strain refer to shear deformation on a primary plane in a particular direction. The single crystal resolution, Eqs. (4.12) and (4.13), already have included cross-sectional area changes in their formulation.

Eqs. (4.21) and (4.22) represent a direct intercomparison of axial nominal, or "engineering" stress-strain behavior in the uniaxial deformation of both the polycrystal and the single crystal. One recalls that the stress specification and the strain measure are, of course, entirely arbitrary. It is not too surprising hence, that the present generalization applies to stress-strain functions which refer to the undeformed state of the material. In the definition of strain measure, it might be convenient to include the temperature dependence $(1 - T/T_m)$ thus referring all strain to the undeformed zero-point state of the crystalline solid.

Considering the finite deformation mode and transition structure in conjunction with the zero-point isotropic moduli distribution of Chapter V and the related multiple isotropic elasticities of Chapter VI, one may conclude that the integers, r and s, are different representations of the same distribution of states in the solid.

To obtain a non-dimensional stress-strain function applicable to all 19 polycrystals and binary combinations, let $\hat{\sigma} = \dfrac{\sigma}{A}$ where $A = 2.89 \times 10^4$ kg/mm^2 is the universal constant for the zero-point isotropic shear modulus of Eq. (5.1). One thus obtains a completely general polycrystalline stress-strain function, Eq. (9.3), for a reference solid.

$$\hat{\sigma} = (2/3)^{s/2 + p/4} B_0 \left(1 - T/T_m\right) \left(\varepsilon - \varepsilon_b\right)^{1/s}. \tag{9.3}$$

Or, for the single crystal, letting $\hat{\tau} = \dfrac{\tau}{A}$, one obtains for Eq. (4.11) a non-dimensional form.

$$\hat{\tau} = (2/3)^{s/2+p/4} \frac{B_0}{(\overline{m})^{3/2}} (1 - T/T_m)(\gamma - \gamma_b)^{1/s}. \qquad (9.4)$$

In both Eqs. (9.3) and (9.4) s is now assumed to refer not only to the index of the stable zero-point isotropic shear modulus of Chapter V, but also to the finite deformation mode index, r.

Eqs. (9.3) and (9.4) state that the finite deformation parabola coefficient in any finite deformation mode is proportional to the zero-point value of whatever multiple isotropic elasticity is applicable. This presumes that in finite distortion all crystalline solids deform under conditions prescribed in a common series of discretely distributed states, not dependent upon crystal structure.

The finite distortional deformation, whether polycrystalline or stage III single crystal, is given solely from a knowledge of s, p, T/T_m, and the strain intercept ε_b or γ_b applicable in the particular deformation mode of interest.

Chapter X

Experiment and Theory

The finite distortional deformation generalization described in this monograph, a generalization which is applicable to both polycrystal and single crystal, is a stage III resolved shear stress, resolved shear strain phenomenon. In proposing atomistic explanations for this behavior considerable care must be exercised when directly or indirectly extrapolating the atomistic concepts and conclusions from stage I or stage II theory and experiment to this stage III region of complex slip which until very recently has received relatively little theoretical attention.

The major topical interest in the mathematical continuum theory of plasticity and in atomistic imperfection theory has been in problems related to the inauguration of plastic deformation: with the nature of yield surfaces; with the quasi-static and dynamic factors influencing elastic limits or critical resolved shear stress; with small strain plastic-elastic fields; and with the dislocation mechanism governing stage I, or "easy glide," deformation. A large number of different atomistic dislocation interaction mechanisms have been proposed to describe the linear stage II work-hardening behavior, Eq. (4.1) or Eq. (4.15). The experimental variability, and consequent theory insensitivity of the measured stage II slope, θ_{II}, and the quantitative approximations in those proposals, have not yet permitted a meaningful comparison of experiment with the many theories predicting linearity in that situation.

From the experimental point of view, the stage III deformation is far better defined for such comparisons, both quantitatively and qualitatively. As has been amply demonstrated in the preceding chapters of this monograph, sufficient polycrystalline and single crystal stage III experimental data have been available to establish the quantitative and qualitative nature of this finite distortion behavior. The critical examination of the applicability of proposed plausible theories is straightforward in this situation for the 27 crystalline solids thus far experimentally studied. Any proposed theory purporting to explain this form of finite distortion must provide stress-strain functions which are parabolic, with a $(1 - T/T_m)$ temperature dependence; parabola coefficients must be independent of specific crystal structure, and they must be proportional to the *zero-point* isotropic linear elastic shear modulus, and they must fall into one of a series of quantized finite deformation modes, with second-order transitions or slope discontinuities from one mode to another at specified critical strains. In addition, any proposed theory must take cognizance of the fact that over a range of nearly 14 orders of magnitude, the basic interrelated stage III stress-strain functions of 24 polycrystalline solids and 12 cubic single crystal solids do *not* depend upon the strain rate, $\dot{\varepsilon}$ or $\dot{\gamma}$.

Four atomistic theories during the past thirty-five years have been developed based upon the foreknowledge of the experimentally observed parabolicity of resolved single crystal stage III deformation. There is (a) the Taylor (1934) dislocation theory which describes workhardening in terms of a uniform distribution of positive and negative dislocations, furnishing Eq. (10.1):

$$\tau = \mu \sqrt{b/L} \, \gamma^{1/2}$$ (10.1)

where μ is the shear modulus at the temperature of the test, b is the BURGER's vector, and L is the distance plastic deformation producing dislocations move before they are inhibited. (b) The Mott (1952) dislocation pile-up theory of workhardening furnishes Eq. (10.2):

$$\tau = \mu \sqrt{\frac{b}{2\pi l}} \, \gamma^{1/2}$$ (10.2)

where l is the length of the Frank-Read source from which dislocations are generated to pile up against a barrier. A constant l is essential for parabolicity. (c) The Fitzgerald (1966) theory in which finite crystal deformation is assumed to arise from particle momentum waves propagating in the periodic crystal lattice, furnishes Eq. (10.3):

$$\beta \text{ (fcc)} = \frac{h}{\pi^2 d_1^3} \left[\frac{f E_1}{O_i \xi_0 m \times 12\sqrt{2}} \right]^{1/2}$$ (10.3)

where β is the parabola coefficient of the final portion of the finite deformation for face-centered cubic crystals; h is PLANCK's constant; O_i is a factor depending upon orientation and the size and shape of the sample; ξ_0 is the initial number of field-free particles per unit length in a particular direction in a crystal lattice; m is the atomic mass; E_1 is an elastic constant in the direction of the closest lattice spacing, d_1; and f is an unknown numerical factor near unity.

(d) The Kuhlmann-Wilsdorf (1967) theory of almost-parabolic stage III deformation which incorporates a dislocation factor with the earlier Kuhlmann-Wilsdorf mesh-length theory of linear stage II workhardening, furnishes Eq. (10.4):

$$\tau = \left[\frac{\mu \sqrt{\varphi b}}{\pi n \sqrt{m f L_c}} \ln \frac{n \sqrt{m f L_c}}{\sqrt{4 \varphi b (\gamma - \gamma_b)}} \right] (\gamma - \gamma_b)^{1/2}$$ (10.4)

where φ (designated as β in the Kuhlmann-Wilsdorf derivation) is a parameter linking the mean free dislocation length to the dislocation density; $n \simeq 3$ is a parameter related to the longest frequent link lengths in the walls of an assumed cell structure; f is the fraction of volume occupied by these cell walls; L_c is a constant lower limit of the mean cell diameter after stage III deformation has begun; and γ_b is the stage III parabola intercept on the strain abscissa (see Fig. 4.1 above).

All of these atomistic proposals contain parameters which, if experiment is to be used to discriminate among the proposed explanations, must be determined from completely independent reasonable estimates, or preferably from completely separate experiment, rather than from the direct or indirect use of the experimental stress-strain functions to be predicted. As the number of qualitative features which such theories include increases, so does the number of assumptions which must be made, and hence the number of additional parameters which must be defined and independently determined.

Before the workhardening phenomenon had received much theoretical attention, the Taylor (1934) and the Mott (1952) theories were developed to provide a plausible explanation of workhardening. Each has a length factor, L or l, of approximately 10^{-4} cm whose precise numerical values never have been independently determinable so that a quantitative distinction between those theories could be made. COTTRELL (1953) discarded both of those proposals, primarily on the basis that they failed to include the then newly discovered stage II linear workhardening. It should be noted, however, that stage III deformation is unaltered when stage II deformation is absent. (BELL, 1964, 1965a; Chapter IV above.)

Both the Fitzgerald (1966) and the Kuhlmann-Wilsdorf (1967) theories provide for a nearly-linear stage II and a nearly-parabolic stage III finite deformation, with 3 to-be-determined parameters, 4 known constants, and 5 to-be-determined parameters, and 2 known constants, respectively. The Fitzgerald theory also provides for a Portevin - le Chatelier phenomenon in a finite lattice, for a low-stress stage I region, and for an approximation of the present writer's Eqs. (2.24) and (2.25). The assumed atomic particle momentum wave furnishes a critical velocity in aluminum and in copper from the de Broglie wave relation in close numerical agreement with one of the experimentally known 5 finite deformation transition velocities for each metal.

The Kuhlmann-Wilsdorf theory provides for a stage II deformation smoothly proceeding into a stage III at a predicted resolved shear stress, τ^*, and specifies that they be related in terms of the present writer's experimentally established interrelation ot the two deformation stages, Eqs. (4.5) and (4.6) from which $\beta_{ro}(II)$ zero-point parabola coefficients were determined (see Appendix I). The prediction of the location of γ_b, as in Eq. (4.6), is of more than small interest since this intercept parameter, γ_b, is one of the 4 fundamental quantities to be specified in the present writer's parabolic generalization.

None of those atomistic proposals explicitly includes temperature dependence other than that which is introduced through the shear modulus or elastic constant and lattice spacing. A large fraction of the quasi-static polycrystalline data of Chapter III was plotted in terms of $\sigma/(1 - T/T_m)$ for values of T/T_m from 0.004 to 0.94 (values of T from 4.2° K to 1,809° K), which, together with the dynamic diffraction grating measurements to $T/T_m = 0.98$ (Fig. 2.21) and the 660 experimentally determined single crystal parabola coefficients shown in Table X, Chapter IV, for T/T_m from 0.003 to 0.937, provide a precise experimental temperature dependence of $(1 - T/T_m)$.

With equal precision there has been established a dependence of parabola coefficients upon the zero-point isotropic shear modulus, $\mu(0)$, and *not* upon the test temperature value, μ or E_1, whose temperature dependence at low temperatures, Eq. (5.4) and Fig. 5.2, and particularly at higher temperatures, are very different from $(1 - T/T_m)$.

This experimental dependence of the finite distortion deformation upon the *zero-point* isotropic shear modulus and not upon that of the test temperature, raises serious problems for all atomistic theories, including the above-mentioned four, which are based upon the behavior of the plastically undeformed surrounding elastic domain. The Fitzgerald (1966) theory, with its dependence upon directional moduli and specific lattice spacing, introduces a dependency of parabola coefficients upon crystal structure, contrary to the present stage III experimental evidence for finite distortion. The Fitzgerald theory also requires that $O_t \xi_o m = $ constant be assumed, since experimental parabola coefficients definitely do not depend upon the mass of the element, the orientation, or the shape and size of the specimen sample.

As an example of a somewhat different atomistic point of view, GILMAN (1965 a, b) proposed a semi-empirical strain rate—stress function based upon the atomistic premise that the strain rate, $\dot{\gamma}$, is given by the BURGER's vector multiplied by the product of the mobile dislocation density and the dislocation velocity. It is a simple matter to show that the resulting viscosity-dependent strain rate—stress function, even with 5 parameters to be determined by auxiliary experiment, cannot be extrapolated (even for a constant strain rate experiment) from stage I or stage II deformation to the stage III parabolic *non-viscous* single crystal and polycrystal finite distortional deformation.

Despite the qualitative parallels that are found in some instances and, in particular, for the Fitzgerald (1966) particle wave hypothesis, the atomistic theories described above are too indeterminate to provide a satisfactory theoretical explanation for the present experimental finite distortion generalization. It is difficult to see where in those proposed explanations the experimentally observed discretely distributed finite deformation mode and transition stability structure can arise. The experimentally observed *independence* of the finite distortion stress-strain function with respect to crystal structure, test temperature elastic constants, mass, lattice spacing, stacking fault energy, and viscosity, is a matter of considerable quantitative precision.

Because of the multiplicity of assumptions and attendant multiple parameters to be determined by auxiliary arguments or experiments in the above-mentioned atomistic theories, and because of the fact that none of those atomistic theories provide for the known $(1 - T/T_m)$ temperature dependence and for the known dependence upon the zero-point moduli instead of the test temperature moduli, the relation between theory and experiment for stage III finite distortional deformation is still far from satisfactory. Each of the experimentally established conditions places considerable restriction upon what may be considered to be plausible initial hypotheses. Finite distortional deformation being independent of specific crystal structure and test temperature elastic constants, for example, emphasizes the importance of a domain structure larger than lattice dimensions in which small lattice distortion provides a series of quantized states.

Unlike the unsatisfactory situation which exists when one attempts to interrelate experiment and atomistic theory for stage III stress-strain functions, one may meaningfully compare non-linear continuum experiment and theory. Introducing the Cristescu (1963a, b; 1965a, b) general stress-strain function, Eq. (10.5), which one may consider in conjunction with the equation of motion, Eq. (10.6), and the continuity equation, Eq. (10.7), to describe one-dimensional finite amplitude wave propagation in solids, we have:

$$\frac{\partial \varepsilon}{\partial t} = \varphi\,(\sigma, \varepsilon)\,\frac{\partial \sigma}{\partial t} + \psi\,(\sigma, \varepsilon) \tag{10.5}$$

$$\varrho\,\frac{\partial v}{\partial t} = \frac{\partial \sigma}{\partial x} \tag{10.6}$$

$$\frac{\partial v}{\partial x} = \frac{\partial \varepsilon}{\partial t}. \tag{10.7}$$

The over 1,000 diffraction grating wave propagation experiments in aluminum, copper, lead, zinc, magnesium, nickel, and α-brass, as was shown in Chapter II, have provided an overwhelming amount of experimental evidence that:

$$\varphi = \frac{2\,(\varepsilon - \varepsilon_b)}{\sigma}$$
$$\psi = 0\,.$$

(10.8)

The *empirical* correlation between this $^1/_2$ secant modulus governing the finite amplitude wave speeds and the transverse wave speeds in TRUESDELL's (1961) theory of waves in finite elasticity has been described elsewhere (BELL, 1962a, 1967a).

In the finite amplitude wave theory of TAYLOR (1942), VON KARMAN (1942), and RAKHMATULIN (1945), $\varphi\,(\sigma,\varepsilon)$ need not be specified *a priori*, and one sets $\psi = 0$ by assuming that $\sigma = \sigma\,(\varepsilon)$. The form of φ in Eq. (10.8) was, of course, obtained experimentally after demonstrating that the finite amplitude wave theory had applied in the various solids (see Chapter II above).

Visco-plasticity analyses have assumed an explicit form for the stress-strain function, Eq. (10.9), and MALVERN (1951) also assumed an alternative possibility, Eq. (10.10).

$$\varphi = 1/E \qquad \psi = k\left[\sigma - \sigma_{st}\,(\varepsilon)\right]$$

(10.9)

$$\varphi = 1/E \qquad \psi = a\,e^{k\,[\sigma - \sigma_n(\varepsilon)]}$$

(10.10)

where E is YOUNG's modulus, k and a are constants to be determined, and $\sigma_{st}(\varepsilon)$ is a quasi-static stress-strain function.

A number of persons have performed computer calculations combining Eqs. (10.6), (10.7), and (10.9) or (10.10) for a variety of assumed initial conditions at the impact face. In numerous papers, independent of such computations, the present writer has shown from experiment that φ could not be of the form of Eq. (10.9). (BELL, 1960b, 1961a, b; 1965b; BELL and STEIN, 1962.) It was shown that the maximum amplitude of the initial small precursor wave which travels with the elastic bar velocity did not exceed the quasi-static elastic limit, as would be expected of finite amplitude wave speeds assuming Eqs. (10.9) or (10.10). LUBLINER (1964) proposed that by an appropriate modification of φ somewhere between Eqs. (10.8) and (10.9), experimental strain-time data might be empirically matched with experiment for ψ small, but not equal to zero, and thus a desired viscous dependence, however small, might be preserved.

In the classical tradition of comparing experiment and theory, the present writer's finite amplitude wave propagation experimental data alone are more than adequate in the several metals studied for establishing that $\psi = 0$. When to these dynamic data are added the experimentally determined stress-strain functions from hundreds of quasi-static experiments for strain rates varying by nearly 14 orders of magnitude for 19 crystalline elements and 8 binary combinations, and the resolved dynamic and quasi-static stage III data of 12 cubic single crystals, the experimental evidence is overwhelming that for the finite distortion of crystalline solids, the stress-strain function does *not* depend explicitly on viscosity, i.e., $\psi = 0$ in Eq. (10.5).

A second correlation between theoretical prediction and experiment is the remarkable quantitative accord in terms of the Taylor (1938) aggregate theory stress and strain ratios, Eq. (4.8), between the resolved stage III shear stress-strain function of cubic crystals when calculated as macroscopic *single* slip, and the polycrystal uniaxial stress-strain function when referred to the *undeformed* state of the solid. Since the theoretical development was for face-centered cubic solids, the applicability of the stress and strain ratios, Eq. (4.8), with exactly the same numerical bound, $\overline{m} = 3.06$, to body-centered cubic solids as has been found here to be the experimental behavior, is curious indeed. The earlier experimental aggregate studies of GREENOUGH (1952), BARRETT and LEVENSON (1940), BATEMAN (1954), BOAS and OGILVIE (1954), and ROSENTHAL and GRUPEN (1962) definitely have demonstrated that, contrary to assumption, aggregate grains deform inhomogeneously; that a large fraction of grains do not rotate as predicted; that face-centered cubic aggregate slip is not necessarily confined to {111} planes as assumed; that surface effects are significant; and that stress distribution in uniaxial deformation is not necessarily uniform in the aggregate.

The remarkable quantitative agreement shown experimentally in Chapter IV above, found when single crystal and polycrystal are compared (including stress-strain functions, critical strains, Portevin-le Chatelier effect detail, and binary combination behavior), suggests that the orientation summation processes are theory insensitive, and are not dependent upon the detailed assumptions from which the theory was developed. An aggregate theory, basically independent of crystal structure, which included a parallel orientation summation process of a macroscopic single slip inhomogenous deformation upon primary planes, and which included crystallographic and rigid body rotations and surface grain discontinuities, and which allowed for inhomogeneous stress among the aggregate components, if developed, should provide the identical stress and strain ratios as are in the Taylor theory. If there were such a development, a complete consistency would be achieved between aggregate theory and experiment as now known.

The phenomenon of finite deformation stability has been shown in this monograph to be a main feature of finite distortional deformation. The experimental evidence shown in Chapters V, VI, and VII strongly suggests that the phenomenon is produced by second-order phase changes arising from the small distortions at relatively high stress of the bulk solid between planes in block shear.

The experimental finite hydrostaticity studies of transition phenomena by P. W. BRIDGMAN and others offer striking parallels with the present distortional deformation transitions, but in the present instance nearly all of the transitions are second-order slope discontinuities. One might loosely introduce the term, "deformation phase," assuming that these finite distortion transitions are second-order phase changes awaiting the development of a theory of finite elasticity which includes such a phase description and is compatible with the GIBBSIAN thermodynamics of phases.

The experimental evidence developed in Chapters V, VI, and VII demonstrating that the finite deformation mode index, r, and the zero-point isotropic shear moduli distribution, s, were different representations of the same phenomenon, has shown that the finite distortional deformation may be expressed in terms of Eqs.(9.3) or (9.4) for a single reference crystalline solid which includes all 27 crystalline solids experimentally studied, solely from a knowledge of the integers, s and p; the fractional melting point temperature, T/T_m; and the intercept of the deformation mode of interest upon the strain abscissa, ε_b.

At the present moment the experimental data described in this monograph suggest strong possibilities for a theory of block shear in terms of macroscopic single slip on primary crystallographic planes, related to zero-point isotropic linear elastic forces which are numerically dependent upon the state of the material. Such a theory would have to include a stability structure arising from a series of quantized material states dependent upon the magnitude of the elastic distortion, undergoing a change from one member of a discrete distribution to another at specified critical shear angles. The initial mode index of the particular state governing the block shear would have to be dependent upon the purity, the prior deformation history, the prior thermal history, etc., as has been described in the first nine chapters of this monograph. Transitions at critical strains, including whether they occur or not and the magnitude of the mode index when they do occur, also would have to be shown to be dependent upon these same parameters.

From over 1,000 finite distortional deformation uniaxial stress experiments described in the first nine chapters of this monograph, the linearly temperature-dependent generalized parabolic stress-strain function is found to be solely dependent upon the quantized zero-point isotropic moduli integers, s and p, the fractional melting-point temperature T/T_m, and the deformation mode strain abscissa intercept ε_b or γ_b. Some of the following parameters may influence the operative value of the integer, s, as has been described in detail in this monograph, but in every instance the generalization still applies, *regardless* of variations in the:

crystal structure;

strain rate (from 0.000000006 sec^{-1} to 70,000.0 sec^{-1});

purity;

prior deformation history;

prior thermal history;

grain size;

geometry of the specimen cross-section;

percentage composition of binary combinations;

specimen diameter (0.075 mm to 25.4 mm diameter solid cylinders, and various thicknesses of thin-walled hollow tubes);

specimen length to diameter ratio (3 to 233);

magnitude of elastic limit, Y (0.050 kg/mm^2 to 80 kg/mm^2);

type of experiment, whether uniaxial compression, uniaxial tension, or the torsion of hollow tubes; whether uniform quasi-static or distributed non-linear wave propagation;

fractional melting point T/T_m, from $T/T_m = 0.003$ to $T/T_m = 0.98$; and

whether or not phenomena such as the Portevin-le Chatelier effect are superimposed upon the basic parabolic stress-strain function.

The experimental fact that a quantized distribution of elastic states presents itself in such a simple form for the asymptotically infinite body situation suggests intriguing new possibilities for the study of limiting processes in non-linear fields. The fact that the non-metallic ionic crystal, NaCl, is included in the present generalization, with stress-strain functions directly determinable from metallic crystals (see Fig.8.6) is of more than minor significance for such studies.

The present monograph has dealt almost entirely with finite distortion stress-strain functions. In a separate, nearly-completed monograph, this generalized parabolic polycrystalline stress-strain function is used to examine the results of several hundred dynamic plasticity experiments. Those studies include three-dimensional finite amplitude wave initiation, growth, and propagation; non-linear finite amplitude wave interaction between loading waves, and between loading and unloading waves; incremental waves in prestressed fields; the phenomenon of reflection and transmission at elastic-plastic and plastic-elastic boundaries, etc.

The simplicity and degree of generality of the unified finite distortional deformation behavior, together with the additional discovery that finite distortional deformation stability properties based upon the existence of hitherto unknown discretely distributed multiple linear elastic states, are important, demonstrates the power of plausible, clearly-defined, and sufficiently simple experiment to establish patterns of understanding which must be included in any theory purporting to deal with the physics of finite distortion in crystals.

Appendix I

T/T_m	Crystal	Nomen-clature	Tem-pera-ture °K	No. of Specs.	Purity	θ_{II} (kg/mm²)	τ^* (kg/mm²)	$\beta_{ro}(II)$ (kg/mm²)	Mode Index r
0.003	Cu	16.5	4.2	1	H	17.8	14.2	22.55	1
0.003	Cu	15.1	4.2	1	H	16.4	14.8	22.10	1
0.003	Cu	15.4	4.2	1	H	15.3	13.2	20.16	1
0.003	Cu	S8.9	4.2	1	H	17.4	14.4	22.46	1
0.003	Cu	S9.0	4.2	1	H	15.7	9.8	17.59	2
0.005	Al	P	4.2	1	H	—	—	13.38	1
0.007	Pb	10a	4.2	1	H	3.52	1.20	2.93	1
0.007	Pb	17a	4.2	1	H	3.75	1.50	3.43	1
0.007	Pb	20	4.2	1	H	3.12	1.68	3.26	1
0.007	Pb	31b	4.2	1	H	3.15	0.80	2.26	3
0.012	Ni	6c	20	1	L	30.3	15.6	31.11	2
0.025	Pb	34b	15	1	H	3.45	1.10	2.82	2
0.033	Pb	17b	20	1	H	3.65	1.32	3.20	1
0.045	Ni	6a	78	1	L	27.8	14.6	29.8	2
0.045	Ta	—	147	1	L	—	—	—	—
0.052	Ni	A7a	90	1	H	26.0	2.21	11.31	7
0.052	Ni	A21	90	1	H	29.2	2.32	12.27	6
0.052	Ni	C2a	90	1	H	27.5	1.96	10.95	7
0.052	Ni	C19	90	1	H	25.0	2.54	11.89	7
0.052	Ni	—	90	1	L	23.68	7.0	19.21	4
0.052	Ni	0—18	90	1	L	23.9	9.2	22.0	4
0.052	Ni	0—19	90	1	L	22.5	9.5	21.7	4
0.054	Ni	—	93	1	L	29.85	5.9	19.83	4
0.057	Cu	—	77	1	H	16.00	11.0	19.90	2
0.057	Cu	—	77	1	H	15.00	10.1	18.45	2
0.057	Cu	16.4	77	1	H	15.9	9.8	18.71	2
0.057	Cu	15.5	77	1	H	14.5	10.0	18.06	2
0.057	Cu	S14.3	77	1	H	15.4	7.4	16.01	3
0.057	Cu	S8.8	77	1	H	16.2	9.2	18.31	2
0.057	Cu	S8.2	77	1	H	14.8	8.3	16.61	2
0.057	Cu	S9.7	77	1	H	15.2	7.8	16.33	3
0.057	Cu	S14.7	77	1	H	15.1	6.8	15.19	3
0.057	Cu	S14.8	77	1	H	14.8	6.7	14.93	3
0.058	Cu	305D	78	1	H	—	—	—	—
0.058	Cu	304C	78	1	H	—	—	—	—
0.060	Ta	—	197	1	L	—	—	—	—
0.067	Au	—	90	1	H	10.00	1.58	6.10	5
0.070	Au	—	93	1	H	9.45	2.1	6.74	4
0.073	Ag	3	93	1	H	8.00	2.35	6.61	4
0.073	Ag	7	93	1	H	12.27	1.18	5.90	5
0.073	Ag	8	93	1	H	8.35	2.08	6.35	5
0.076	Ta	—	247	1	L	—	—	—	—
0.083	Ta	—	272	1	L	—	—	—	—
0.083	Cu	—	113	1	H	15.00	9.2	18.10	2

$\beta_{ro}(\mathrm{III})$ (kg/mm²)	Mode Index r	1st Trans. $\beta_{ro}(\mathrm{III})$ (kg/mm²)	Mode Index r	2nd Trans. $\beta_{ro}(\mathrm{III})$ (kg/mm²)	Mode Index r	References
—	—					Mitchell and Thornton (1963)
—	—					Mitchell and Thornton (1963)
—	—					Mitchell and Thornton (1963)
—	—					Mitchell and Thornton (1963)
—	—					Mitchell and Thornton (1963)
12.78	1					Noggle and Koehler (1957)
—	—					Bolling, Hays and Wiedersich (1962)
—	—					Bolling, Hays and Wiedersich (1962)
—	—					Bolling, Hays and Wiedersich (1962)
—	—					Bolling, Hays and Wiedersich (1962)
27.72	2					Haasen (1958)
—	—					Bolling et al. (1962)
—	—					Bolling et al. (1962)
24.2	3					Haasen (1958)
25.38	2					Mitchell and Spitzig (1965)
—	—					Mader, Seeger and Leitz (1963)
—	—					Mader, Seeger and Leitz (1963)
—	—					Mader, Seeger and Leitz (1963)
—	—					Mader, Seeger and Leitz (1963)
—	—					Kronmüller (1959)
—	—					Meissner (1959)
—	—					Meissner (1959)
—	—					Andrade and Henderson (1951)
18.66	2					Thornton, Mitchell and Hirsch (1962)
17.50	2					Thornton, Mitchell and Hirsch (1962)
—	—					Mitchell and Thornton (1963)
—	—					Mitchell and Thornton (1963)
—	—					Mitchell and Thornton (1963)
—	—					Mitchell and Thornton (1963)
—	—					Mitchell and Thornton (1963)
—	—					Mitchell and Thornton (1963)
—	—					Mitchell and Thornton (1963)
—	—					Mitchell and Thornton (1963)
12.83	4					Blewitt and Coltman (1955)
10.16	5					Blewitt and Coltman (1955)
26.05	2	16.82	4			Mitchell and Spitzig (1965)
6.61	4					Berner (1960)
8.32	3					Andrade and Henderson (1951)
6.24	5	8.68	3			Andrade and Henderson (1951)
—	—					Andrade and Henderson (1951)
—	—					Andrade and Henderson (1951)
17.23	4	11.76	6			Mitchell and Spitzig (1965)
14.28	5	5.69	9			Mitchell and Spitzig (1965)
16.14	3					Thornton et al. (1962)

Appendix I (continued)

T/T_m	Crystal	Nomen-clature	Tempera-ture °K	No. of Specs.	Purity	θ_{II} (kg/mm²)	τ^* (kg/mm²)	$\beta_{ro}(II)$ (kg/mm²)	Mode Index r
0.083	Cu	S8.7	113	1	H	15.50	6.8	15.84	3
0.084	Al	—	78	1	H	—	—	—	—
0.084	Al	DD	78	1	H	9.36	1.31	5.40	5
0.084	Al	15	78	1	H	14.1	1.25	6.48	5
0.091	Ta	—	296	1	L	2.15	6.9	6.00	9
0.094	Al	—	88	8	L	—	—	—	—
0.097	Al	—	90	1	H	—	—	—	—
0.097	Al	5	90	1	H	7.22	0.90	4.04	7
0.097	Al	1	90	1	H	11.05	1.48	6.28	5
0.097	Al	8	90	1	H	13.89	1.19	6.38	5
0.099	Ta	—	323	1	L	4.80	6.35	8.67	7
0.100	Cu	S14.2	135	1	H	13.7	6.3	14.59	3
0.106	Ta	—	348	1	L	9.80	5.30	11.40	6
0.112	Ag	2	193	1	H	7.56	1.38	5.42	5
0.112	Ag	4	193	1	H	7.89	1.48	5.73	5
0.112	Ag	9	193	1	H	13.51	1.25	6.90	4
0.112	Ag	10	193	1	H	8.25	2.79	8.06	3
0.113	Cu	16.6	153	1	H	14.0	6.5	15.21	3
0.113	Ni	A5a	195	1	H	24.7	1.92	10.97	7
0.113	Ni	B1	195	1	H	24.1	1.90	10.77	7
0.113	Ni	C5a	195	1	H	22.4	3.1	13.28	6
0.114	Ta	—	373	1	L	10.60	4.95	11.56	6
0.114	Ta	9.5×10^{-1}	373	1	L	—	—	—	—
0.114	Ta	9.5×10^{-2}	373	1	L	4.05	6.10	7.94	8
0.114	Ta	4.8×10^{-2}	373	1	L	4.33	5.70	7.94	8
0.114	Ta	9.3×10^{-3}	373	1	L	7.15	5.65	10.15	6
0.114	Ta	9.3×10^{-4}	373	1	L	10.60	4.95	11.56	6
0.114	Ta	9.3×10^{-5}	373	1	L	12.8	4.35	11.91	6
0.114	Ta	9.3×10^{-6}	373	1	L	14.0	3.90	11.80	6
0.114	Ta	14	373	1	L	17.40	6.40	16.86	4
0.114	Ta	30	373	1	L	8.57	4.65	10.08	6
0.114	Ta	29	373	1	L	8.80	4.70	10.27	6
0.114	Ta	28	373	1	L	5.96	5.55	9.18	7
0.114	Ta	31	373	1	L	9.63	4.55	10.57	6
0.114	Ta	27	373	1	L	8.96	4.40	10.03	6
0.114	Ta	8	373	1	L	9.80	3.90	9.87	6
0.114	Ta	15	373	1	L	7.58	6.10	10.86	6
0.114	Ta	19	373	1	L	10.60	5.20	11.85	6
0.122	Ta	—	398	1	L	11.85	4.65	11.96	6
0.127	Cu	S14.5	172	1	H	13.1	5.6	13.87	3
0.128	Pb	24a	77	1	H	2.48	0.45	1.71	4
0.128	Pb	24b	77	1	H	2.61	0.395	1.65	4
0.128	Pb	25a	77	1	H	2.60	0.25	1.31	5
0.128	Pb	25b	77	1	H	2.60	0.345	1.54	5
0.128	Pb	27a	77	1	H	2.63	0.275	1.38	5
0.128	Pb	27b	77	1	H	2.90	0.335	1.59	5

β_{ro}(III) (kg/mm^2)	Mode Index r	1st Trans. β_{ro}(III) (kg/mm^2)	Mode Index r	2nd Trans. β_{ro}(III) (kg/mm^2)	Mode Index r	References
—	—					MITCHELL and THORNTON (1963)
4.75	6					BERNER (1960)
7.11	4					NOGGLE and KOEHLER (1957)
6.77	4					NOGGLE and KOEHLER (1957)
5.80	9					MITCHELL and SPITZIG (1965)
8.05	3	5.355	6			BOAS and SCHMID (1931)
4.94	6					BERNER (1960)
4.94	6					KELLY (1956)
6.19	5					KELLY (1956)
6.77	4					KELLY (1956)
8.19	7	5.23	10			MITCHELL and SPITZIG (1965)
—	—					MITCHELL and THORNTON (1963)
10.47	6	5.43	9			MITCHELL and SPITZIG (1965)
5.72	5	7.95	4			ANDRADE and HENDERSON (1951)
6.24	5	7.11	4	2.345	10	ANDRADE and HENDERSON (1951)
8.72	3	6.32	5			ANDRADE and HENDERSON (1951)
8.38	3					ANDRADE and HENDERSON (1951)
—	—					MITCHELL and THORNTON (1963)
—	—					MADER et al. (1963)
—	—					MADER et al. (1963)
—	—					MADER et al. (1963)
11.10	6	7.98	8			MITCHELL and SPITZIG (1965)
8.96	7					MITCHELL and SPITZIG (1965)
7.09	8	6.86	8			MITCHELL and SPITZIG (1965)
7.58	8	5.11	10			MITCHELL and SPITZIG (1965)
9.44	7	5.91	9			MITCHELL and SPITZIG (1965)
9.32	7					MITCHELL and SPITZIG (1965)
10.30	6					MITCHELL and SPITZIG (1965)
10.21	6					MITCHELL and SPITZIG (1965)
15.64	4					MITCHELL and SPITZIG (1965)
9.13	7	6.43	9			MITCHELL and SPITZIG (1965)
9.58	7	7.71	8			MITCHELL and SPITZIG (1965)
8.15	7					MITCHELL and SPITZIG (1965)
9.03	7					MITCHELL and SPITZIG (1965)
8.76	7					MITCHELL and SPITZIG (1965)
8.74	7	4.99	10			MITCHELL and SPITZIG (1965)
10.10	6					MITCHELL and SPITZIG (1965)
10.10	6					MITCHELL and SPITZIG (1965)
10.90	6	8.26	7			MITCHELL and SPITZIG (1965)
—	—					MITCHELL and THORNTON (1963)
—	—					BOLLING et al. (1962)
—	—					BOLLING et al. (1962)
—	—					BOLLING et al. (1962)
—	—					BOLLING et al. (1962)
—	—					BOLLING et al. (1962)
—	—					BOLLING et al. (1962)

Appendix I (continued)

T/T_m	Crystal	Nomen-clature	Tem-pera-ture °K	No. of Specs.	Purity	θ_{II} (kg/mm²)	τ^* (kg/mm²)	$\beta_{ro}(II)$ (kg/mm²)	Mode Index r
0.128	Pb	28a	77	1	H	3.45	0.33	1.73	4
0.128	Pb	28b	77	1	H	3.25	0.28	1.55	5
0.128	Pb	29a	77	1	H	3.30	0.30	1.62	4
0.128	Pb	31a	77	1	H	2.60	0.33	1.50	5
0.129	Cu	—	175	1	H	15.00	8.1	17.90	2
0.129	Cu	S8.3	175	1	H	14.9	5.6	14.83	3
0.145	Au	—	193	1	H	10.12	1.53	6.52	5
0.145	Ta	—	473	1	L	11.50	3.82	10.96	6
0.147	Cu	305E	200	1	H	—	—	—	—
0.147	Cu	304A	200	1	H	—	—	—	—
0.150	Pb	13a	90	1	H	2.95	0.66	2.32	3
0.162	Cu	16.8	219	1	H	13.3	4.6	13.19	4
0.162	Fe	366	293	1	L	—	—	—	—
0.162	Fe	62	293	1	L	14.25	6.22	15.95	5
0.162	Fe	3C	293	1	L	—	—	—	—
0.162	Fe	6C	293	1	L	—	—	—	—
0.162	Fe	7C	293	1	L	—	—	—	—
0.162	Fe	4	293	1	L	—	—	—	—
0.164	Al	—	153	1	H	—	—	—	—
0.168	Cu	S8.4	228	1	H	15.6	5.7	16.04	3
0.168	Cu	—	228	1	H	15.0	7.4	17.95	2
0.168	Ni	—	290	1	L	17.30	7.3	19.09	4
0.169	Ni	A18	292	1	H	27.5	1.08	9.28	8
0.169	Ni	C21	292	1	H	23.0	2.0	11.54	7
0.170	Ni	A20	293	1	H	25.0	2.15	12.48	6
0.170	Ni	A27	293	1	H	25.0	1.44	10.22	7
0.170	Ni	C17	293	1	H	27.0	1.67	11.43	7
0.170	Ni	—	293	1	L	22.99	4.35	17.03	5
0.170	Ni	H13b	293	1	L	19.0	4.0	14.85	5
0.170	Ni	H16a	293	1	L	19.5	4.4	15.78	5
0.170	Ni	H17c	293	1	L	19.0	3.0	12.87	6
0.171	Ni	0—7	295	1	L	23.5	2.32	12.60	6
0.171	Ni	0—17	295	1	L	22.3	1.56	11.40	7
0.173	Ni	D1	298	1	H	25.4	2.2	12.78	6
0.174	Ni	6b	300	1	L	20.6	5.7	18.56	4
0.175	Ta	—	573	1	L	9.50	3.10	9.30	7
0.176	Cu	S14.4	238	1	H	14.4	4.3	13.50	3
0.195	Pb	9b	117	1	H	2.55	0.42	1.81	4
0.213	Ni	C2b	368	1	H	21.5	1.8	11.19	7
0.216	Cu	—	293	1	L	—	—	—	—
0.216	Cu	—	293	1	H	—	—	—	—
0.216	Cu	C14/11	293	1	H	13.50	2.65	10.75	5
0.216	Cu	C17/2	293	1	H	15.1	3.30	12.71	4
0.216	Cu	C18/1	293	1	H	13.70	3.29	12.05	4
0.216	Cu	C21/0	293	1	H	15.0	3.02	12.10	4
0.216	Cu	C25/0	293	1	H	13.50	2.98	11.31	4

$\beta_{ro}(III)$ (kg/mm²)	Mode Index r	1st Trans. $\beta_{ro}(III)$ (kg/mm²)	Mode Index r	2nd Trans. $\beta_{ro}(III)$ (kg/mm²)	Mode Index r	References
—	—					Bolling et al. (1962)
—	—					Bolling et al. (1962)
—	—					Bolling et al. (1962)
—	—					Bolling et al. (1962)
16.53	2					Thornton et al. (1962)
—	—					Mitchell and Thornton (1963)
6.40	5					Andrade and Henderson (1951)
9.69	7	7.22	8			Mitchell and Spitzig (1965)
10.96	4					Blewitt et al. (1955)
12.32	4					Blewitt et al. (1955)
—	—					Bolling et al. (1962)
—	—					Mitchell and Thornton (1963)
13.30	6					Dohi (1960)
16.13	5	13.20	6			Dohi (1960)
27.34	2	17.33	5			Taylor (1934)
27.34	2	19.05	4			Taylor (1934)
27.34	2	19.05	4			Taylor (1934)
29.81	2	20.31	4			Taylor (1934)
3.39	8					Berner (1960)
—	—					Mitchell and Thornton (1963)
16.00	3					Thornton et al. (1962)
—	—					Andrade and Henderson (1951)
—	—					Mader et al. (1963)
—	—					Mader et al. (1963)
—	—					Mader et al. (1963)
—	—					Mader et al. (1963)
—	—					Mader et al. (1963)
—	—					Kronmüller (1959)
—	—					Mader et al. (1963)
—	—					Mader et al. (1963)
—	—					Mader et al. (1963)
—	—					Meissner (1959)
—	—					Meissner (1959)
—	—					Mader et al. (1963)
16.95	5					Haasen (1958)
8.24	7					Mitchell and Spitzig (1965)
—	—					Mitchell and Thornton (1963)
—	—					Bolling et al. (1962)
—	—					Mader et al. (1963)
11.34	4	8.12	6			Taylor and Elam (1925)
11.48	4					Mitchell and Thornton (1963)
11.05	4					Diehl (1956)
12.82	4					Diehl (1956)
11.46	4					Diehl (1956)
12.17	4					Diehl (1956)
12.82	4					Diehl (1956)

Appendix I (continued)

T/T_m	Crystal	Nomen-clature	Tem-pera-ture °K	No. of Specs.	Purity	θ_{II} (kg/mm²)	τ^* (kg/mm²)	$\beta_{ro}(II)$ (kg/mm²)	Mode Index r
0.216	Cu	C26/0	293	1	H	13.80	2.28	10.01	5
0.216	Cu	C27/2	293	1	H	15.7	3.95	14.27	3
0.216	Cu	C28/0	293	1	H	14.70	2.55	10.90	4
0.216	Cu	A	293	1	H	12.92	3.00	11.15	4
0.216	Cu	E	293	1	H	17.48	2.63	12.29	4
0.216	Cu	F	293	1	H	12.56	2.48	10.10	5
0.217	Au	—	290	1	H	10.40	0.75	5.04	6
0.217	Cu	—	295	1	H	12.14	2.83	10.60	5
0.217	Cu	—	295	1	H	14.53	2.50	10.91	4
0.217	Cu	—	295	1	H	12.89	2.36	9.97	5
0.217	Cu	—	295	1	H	14.33	2.09	9.90	5
0.217	Cu	—	295	1	H	14.97	1.56	8.73	6
0.217	Cu	—	295	1	H	13.11	2.90	11.12	4
0.217	Cu	—	295	1	H	14.8	6.3	17.50	2
0.217	Cu	—	295	1	L	—	—	—	—
0.217	Cu	16.1	295	1	H	13.4	2.4	10.25	5
0.217	Cu	17.6	295	1	H	11.4	2.9	10.39	5
0.217	Cu	17.2	295	1	H	11.4	2.3	9.25	5
0.217	Cu	17.4	295	1	H	11.9	2.2	9.25	5
0.217	Cu	17.1	295	1	H	11.4	2.1	8.84	6
0.217	Cu	17.3	295	1	H	11.9	1.9	8.59	6
0.217	Cu	17.5	295	1	H	13.0	1.6	8.24	6
0.217	Cu	15.2	295	1	H	13.1	2.7	10.75	5
0.217	Cu	S14.6	295	1	H	15.4	3.6	13.46	3
0.217	Cu	S8.1	295	1	H	14.4	5.2	15.64	3
0.220	Au	—	293	1	L	—	—	—	—
0.2205	Cu	15	299	1	H	12.5	2.33	9.81	5
0.2205	Cu	20	299	1	H	15.5	2.81	12.00	4
0.2205	Cu	24	299	1	H	13.0	2.33	9.97	5
0.2205	Cu	25	299	1	H	13.5	2.90	11.36	4
0.2205	Cu	30	299	1	H	16.3	2.26	11.00	4
0.2205	Cu	32	299	1	H	13.0	2.38	10.09	5
0.2205	Cu	34	299	1	H	17.0	2.25	10.92	4
0.2205	Cu	53	299	1	H	13.5	2.50	10.54	5
0.2205	Cu	65	299	1	H	13.0	2.70	10.74	5
0.2205	Cu	86	299	1	H	11.5	3.15	10.92	4
0.2205	Cu	87	299	1	H	13.8	4.22	13.83	3
0.2205	Cu	88a	299	1	H	10.8	3.40	11.00	4
0.2205	Cu	88b	299	1	H	11.0	2.79	10.06	5
0.2205	Cu	95b	299	1	H	17.5	2.82	12.74	4
0.2205	Cu	128d	299	1	H	11.0	4.28	12.46	4
0.2205	Cu	128f	299	1	H	11.8	3.09	10.99	4
0.221	Cu	305A	300	1	H	—	—	—	—
0.221	Cu	304D	300	1	H	13.95	2.60	10.91	4
0.221	Au	—	295	1	H	10.00	1.30	6.52	5
0.225	Au	1	300	1	L	9.68	0.88	5.32	6

β_{ro}(III) (kg/mm²)	Mode Index r	1st Trans. β_{ro}(III) (kg/mm²)	Mode Index r	2nd Trans. β_{ro}(III) (kg/mm²)	Mode Index r	References
11.28	4					DIEHL (1956)
15.28	3					DIEHL (1956)
12.69	4					DIEHL (1956)
11.59	4					SEEGER et al. (1957)
12.38	4					SEEGER et al. (1957)
9.06	5					SEEGER et al. (1957)
6.39	5					ANDRADE and HENDERSON (1951)
10.73	5					BERNER (1960)
11.68	4					BERNER (1960)
10.12	5					BERNER (1960)
10.11	5					BERNER (1960)
9.41	5					BERNER (1960)
11.65	4					BERNER (1957)
16.11	3					THORNTON et al. (1962)
12.31	4					SACHS and WEERTS (1930)
						MITCHELL and THORNTON (1963)
						MITCHELL and THORNTON (1963)
						MITCHELL and THORNTON (1963)
						MITCHELL and THORNTON (1963)
						MITCHELL and THORNTON (1963)
						MITCHELL and THORNTON (1963)
						MITCHELL and THORNTON (1963)
						MITCHELL and THORNTON (1963)
						MITCHELL and THORNTON (1963)
						MITCHELL and THORNTON (1963)
6.22	5	2.62	9			TAYLOR and ELAM (1925)
11.10	4					SUZUKI et al. (1956)
10.96	4					SUZUKI et al. (1956)
11.99	4					SUZUKI et al. (1956)
12.24	4					SUZUKI et al. (1956)
15.71	3					SUZUKI et al. (1956)
11.10	4					SUZUKI et al. (1956)
12.82	4					SUZUKI et al. (1956)
11.06	4					SUZUKI et al. (1956)
12.17	4					SUZUKI et al. (1956)
11.99	4					SUZUKI et al. (1956)
12.82	4					SUZUKI et al. (1956)
11.10	4					SUZUKI et al. (1956)
11.10	4					SUZUKI et al. (1956)
12.50	4					SUZUKI et al. (1956)
11.69	4					SUZUKI et al. (1956)
11.69	4					SUZUKI et al. (1956)
11.78	4					BLEWITT et al. (1955)
11.12	4					BLEWITT et al. (1955)
6.64	4					BERNER (1960)
6.01	5					SACHS and WEERTS (1930)

Appendix I (continued)

T/T_m	Crystal	Nomen-clature	Tem-pera-ture °K	No. of Specs.	Purity	θ_{11} (kg/mm²)	τ^* (kg/mm²)	$\beta_{ro}(II)$ (kg/mm²)	Mode Index r
0.225	Au	2	300	1	L	11.11	1.0	6.08	5
0.225	Au	3	300	1	L	10.53	0.86	5.48	5
0.225	Au	5	300	1	L	—	—	—	—
0.225	Au	6	300	1	L	10.00	1.05	5.91	5
0.228	Ag	5	290	1	H	9.38	1.0	5.67	5
0.228	Ag	6	290	1	H	8.40	2.15	7.87	4
0.228	Ag	11	290	1	H	8.23	1.10	5.63	5
0.230	Ag	18	293	1	H	10.31	1.45	7.15	4
0.235	Ag	47	300	1	L	7.62	2.68	8.33	3
0.235	Ag	38	300	1	L	8.08	2.92	9.08	3
0.235	Ag	17	300	1	L	8.79	2.0	7.82	4
0.240	Pb	9a	144	1	H	2.45	0.378	1.78	4
0.274	Ni	C5b	473	1	H	23.1	1.1	9.81	8
0.274	Cu	—	371	1	H	14.5	4.4	15.55	3
0.275	Cu	16.2	373	1	H	13.5	1.7	9.34	5
0.275	Cu	S14.9	373	1	H	16.1	2.6	12.62	4
0.275	Cu	S8.5	373	1	H	15.6	2.7	12.66	4
0.283	Pb	13b	170	1	H	2.60	0.42	2.06	3
0.293	Pb	30a	176	1	H	2.94	0.175	1.43	5
0.293	Pb	30b	176	1	H	3.34	0.225	1.72	4
0.293	Pb	32a	176	1	H	2.50	0.20	1.41	5
0.293	Pb	32b	176	1	H	2.80	0.238	1.63	4
0.293	Pb	34a	176	1	H	2.60	0.152	1.25	6
0.293	Pb	33b	176	1	H	3.78	0.138	1.44	5
0.294	Ni	—	508	1	L	13.5	3.5	13.87	6
0.312	Al	—	291	7	L	—	—	—	—
0.314	Al	39T	293	1	L	—	—	—	—
0.314	Al	41T	293	3	L	—	—	—	—
0.314	Al	42T	293	1	L	—	—	—	—
0.314	Al	44T	293	3	L	—	—	—	—
0.314	Al	59.9	293	1	L	—	—	—	—
0.314	Al	72	293	1	L	—	—	—	—
0.314	Al	61.17	293	1	L	—	—	—	—
0.314	Al	3R	293	1	H	11.45	0.37	4.35	7
0.314	Al	515R	293	1	H	11.54	0.62	5.68	5
0.314	Al	613Rb	293	1	H	12.50	0.52	5.41	5
0.314	Al	1R	293	1	H	8.02	0.53	4.28	7
0.314	Al	2R	293	1	H	5.34	0.52	3.53	8
0.314	Al	7R	293	1	H	4.79	0.46	3.12	8
0.314	Al	10R	293	1	H	3.62	0.37	2.43	9
0.314	Al	15R	293	1	H	2.72	0.47	2.30	10
0.314	Al	23R	293	1	H	3.76	0.40	2.55	9
0.314	Al	50Ra	293	1	H	8.52	0.42	3.93	7
0.317	Al	2b	295	1	H	—	—	—	—
0.317	Al	L-12	295	2	H	—	—	—	—
0.317	Al	L-15	295	2	H	—	—	—	—

β_{ro}(III) (kg/mm²)	Mode Index r	1st Trans. β_{ro}(III) (kg/mm²)	Mode Index r	2nd Trans. β_{ro}(III) (kg/mm²)	Mode Index r	References
6.62	4	5.245	6			Sachs and Weerts (1930)
6.65	4	5.83	5			Sachs and Weerts (1930)
6.63	4					Sachs and Weerts (1930)
6.15	5					Sachs and Weerts (1930)
6.89	4	2.60	9			Andrade and Henderson (1951)
7.36	4	5.15	6	2.10	10	Andrade and Henderson (1951)
—	—					Andrade and Henderson (1951)
8.57	3					Andrade and Henderson (1951)
—	—					Sachs and Weerts (1930)
—	—					Sachs and Weerts (1930)
—	—					Sachs and Weerts (1930)
—	—					Bolling et al. (1962)
—	—					Mader et al. (1963)
14.76	3					Thornton et al. (1962)
—	—					Mitchell and Thornton (1963)
—	—					Mitchell and Thornton (1963)
—	—					Mitchell and Thornton (1963)
—	—					Bolling et al. (1962)
—	—					Bolling et al. (1962)
—	—					Bolling et al. (1962)
—	—					Bolling et al. (1962)
—	—					Bolling et al. (1962)
—	—					Bolling et al. (1962)
—	—					Bolling et al. (1962)
13.09	6	9.20	8	4.91	11	Andrade and Henderson (1951)
6.21	5	3.94	7	3.84	7	Boas and Schmid (1931)
5.52	5	4.15	7			Lücke and Lange (1952)
5.38	5					Lücke and Lange (1952)
4.90	6	3.80	7			Lücke and Lange (1952)
5.19	6					Lücke and Lange (1952)
5.09	6					Taylor and Elam (1925)
5.26	6					Taylor and Elam (1925)
5.63	5					Taylor and Elam (1925)
4.31	7	3.01	8			Lücke and Lange (1952)
3.76	7	2.11	10			Lücke and Lange (1952)
5.04	6	2.08	10			Lücke and Lange (1952)
3.57	8	2.36	10			Lücke and Lange (1952)
2.38	—					Lücke and Lange (1952)
2.77	9	2.33	10			Lücke and Lange (1952)
2.30	10					Lücke and Lange (1952)
1.66	11					Lücke and Lange (1952)
2.59	9	1.865	11			Lücke and Lange (1952)
2.26	10					Lücke and Lange (1952)
6.80	4					Staubwasser (1954)
6.37	5					Kingman, Green and Pond (1963)
4.40	6	7.67	4			Kingman, Green and Pond (1963)

Appendix I (continued)

T/T_m	Crystal	Nomen-clature	Tem-pera-ture °K	No. of Specs.	Purity	θ_{II} (kg/mm²)	τ^* (kg/mm²)	$\beta_{ro}(II)$ (kg/mm²)	Mode Index r
0.317	Al	X-10	295	1	H	—	—	—	—
0.317	Al	X-11	295	1	H	—	—	—	—
0.317	Al	X-21	295	1	H	—	—	—	—
0.317	Al	f	295	6	H	—	—	—	—
0.317	Al	h	295	6	H	—	—	—	—
0.317	Al	3	295	5	H	—	—	—	—
0.317	Al	6	295	6	H	—	—	—	—
0.317	Al	3	295	7	H	—	—	—	—
0.317	Al	6	295	9	H	—	—	—	—
0.317	Al	II	295	7	H	—	—	—	—
0.317	Al	—	295	1	H	—	—	—	—
0.317	Al	S	295	1	H	6.58	0.36	3.19	8
0.322	Al	—	300	1	H	—	—	—	—
0.324	Cu	15.6	440	1	H	12.8	1.9	10.32	5
0.324	Cu	15.3	440	1	H	12.0	1.6	9.17	5
0.325	Pb	29b	195	1	H	2.10	0.195	1.34	5
0.349	Cu	—	473	1	H	12.5	3.3	13.95	3
0.349	Cu	S14.10	473	1	H	14.6	2.1	12.02	4
0.349	Cu	S8.6	473	1	H	13.6	1.8	10.75	5
0.349	Cu	S8.10	473	1	H	13.4	1.7	10.37	5
0.359	Ni	A16a	620	1	H	20.9	0.82	9.13	8
0.359	Ni	A17a	620	1	H	24.8	0.8	9.82	8
0.359	Ni	C16	620	1	H	26.5	0.76	10.21	7
0.369	Cu	16.3	500	1	H	13.6	1.0	8.27	6
0.382	Au	—	509	1	H	9.44	0.53	5.65	5
0.400	Al	—	373	10	L	—	—	—	—
0.446	Ni	A16b	770	1	H	23.0	0.78	10.82	7
0.446	Ni	A17b	770	1	H	22.0	0.8	10.71	7
0.452	Au	—	603	1	H	8.23	0.85	6.81	4
0.455	Pb	10b	273	1	H	2.00	0.166	1.48	5
0.459	Cu	608	623	1	H	14.7	0.71	8.50	6
0.483	Ni	—	833	1	L	15.8	2.85	18.30	4
0.489	Ag	12	623	1	H	7.41	0.7	6.66	4
0.500	Au	—	667	1	H	9.41	0.37	5.28	6
0.508	Al	—	473	9	L	—	—	—	—
0.615*	Al	—	573	9	L	—	—	—	—
0.722	Al	—	673	8	L	—	—	—	—
0.768	Cu	608	1043	1	H	2.45	0.27	5.04	8
0.790	Cu	608	1073	1	H	2.02	0.12	3.37	10
0.829	Al	—	773	9	L	—	—	—	—
0.864	Cu	608	1173	1	H	1.60	0.17	5.41	8
0.937	Al	—	873	8	L	—	—	—	—

*	3rd Trans. $\beta_{ro}(III)$ (kg/mm²)	Mode Index r
	2.02	10

$\beta_{ro}(III)$ (kg/mm²)	Mode Index r	1st Trans. $\beta_{ro}(III)$ (kg/mm²)	Mode Index r	2nd Trans. $\beta_{ro}(III)$ (kg/mm²)	Mode Index r	References
4.15	7					KINGMAN, GREEN and POND (1963)
7.12	4					KINGMAN, GREEN and POND (1963)
7.57	4			/		KINGMAN, GREEN and POND (1963)
5.66	5					KINGMAN, GREEN and POND (1963)
5.75	5	4.89	6			KINGMAN, GREEN and POND (1963)
5.89	5					KINGMAN, GREEN and POND (1963)
4.20	7					KINGMAN, GREEN and POND (1963)
6.78	4					KINGMAN, GREEN and POND (1963)
4.50	6					KINGMAN, GREEN and POND (1963)
5.72	5	4.16	7			KINGMAN, GREEN and POND (1963)
2.81	9					BERNER (1960)
2.72	9					NOGGLE and KOEHLER (1957)
10.15	2					POND and HARRISON (1958)
—	—					MITCHELL and THORNTON (1963)
—	—					MITCHELL and THORNTON (1963)
—	—					BOLLING et al. (1962)
14.88	3					THORNTON et al. (1962)
—	—					MITCHELL and THORNTON (1963)
—	—					MITCHELL and THORNTON (1963)
—	—					MITCHELL and THORNTON (1963)
—	—					MADER et al. (1963)
—	—					MADER et al. (1963)
—	—					MADER et al. (1963)
—	—					MITCHELL and THORNTON (1963)
6.22	5					BERNER (1960)
5.80	5	4.002	7	2.91	9	BOAS and SCHMID (1931)
—	—					MADER et al. (1963)
—	—					MADER et al. (1963)
5.76	5	4.28	7	3.46	8	ANDRADE and HENDERSON (1951)
—	—					BOLLING et al. (1962)
7.94	6					ANDRADE and ABOAV (1957)
16.16	5	10.13	7	7.29	9	ANDRADE and HENDERSON (1951)
7.98	4	5.35	6			ANDRADE and HENDERSON (1951)
5.20	6					BERNER (1960)
5.63	5	3.51	8	2.99	8	BOAS and SCHMID (1931)
6.53	5	4.42	6	2.965	8	BOAS and SCHMID (1931)
5.07	6	2.44	9	1.775	11	BOAS and SCHMID (1931)
5.11	8	2.26	12			ANDRADE and ABOAV (1957)
3.09	11	2.34	12			ANDRADE and ABOAV (1957)
5.56	5	2.35	10			BOAS and SCHMID (1931)
5.33	8					ANDRADE and ABOAV (1957)
6.94	4					BOAS and SCHMID (1931)

Appendix II   Table A

Crystal-line Solid	T_m °K	No. of Measurements	Averaged Experimental $\mu(0)$ kg/mm²	Predicted $\mu(0)$ kg/mm² (Eq. 5.1)	s	p	Reference
Os	3,273	1	24,300	23,600	1	0	a
Ir	2,728	1	23,100	23,600	1	0	a
Re	3,443	1	21,400	21,300	1	1	a
Ru	2,773	1	19,500	19,300	2	0	a
W	3,683	1	15,700	15,730	3	0	a
Be	1,550	3	16,000	15,730	3	0	a, b, c
Mo	2,893	1	15,400	15,730	3	0	e
Fe	1,809	4	8,700	8,580	6	0	a, b, c, d
Ni	1,726	2	8,600	8,580	6	0	a, b
U	1,405	2	8,440	8,580	6	0	a, b
Co	1,768	1	7,900	7,750	6	1	d
Ge	1,231	1	7,070	7,010	7	0	h
Ta	3,269	1	7,040	7,010	7	0	f
Pt	2,042	2	5,860	5,720	8	0	a, b
V	2,133	2	5,320	5,160	8	1	a, b
Cu	1,356	4	5,080	5,160	8	1	a, b, c, d
Pd	1,825	1	5,200	5,160	8	1	a
Zn	692.5	4	4,570	4,660	9	0	a, b, c, d
70—30 α-Brass	1,188	1	4,700	4,660	9	0	c
Ti	1,943	2	4,420	4,210	9	1	a, b
Nb	2,743	2	3,900	3,810	10	0	a, b
Zr	2,125	2	3,680	3,810	10	0	a, b
Hf	2,523	1	3,600	3,450	10	1	a
Ag	1,233.8	3	3,170	3,110	11	0	a, b, c
Au	1,336	4	3,090	3,110	11	0	a, b, c, d
Al	933	14	3,110	3,110	11	0	a, b, c, d, g
Th	2,023	2	2,960	3,110	11	0	a, b
Tb	600	1	3,020	3,110	11	0	f
Er	1,798	1	3,180	3,110	11	0	f
Y	1,763	1	2,830	2,810	11	1	f
Dy	1,653	1	2,770	2,810	11	1	f
Ho	1,461	1	2,840	2,810	11	1	f

Appendix II (continued)   Table A (continued)

Crystalline Solid	T_m °K	No. of Measurements	Averaged Experimental $\mu(0)$ kg/mm²	Predicted $\mu(0)$ kg/mm² (Eq. 5.1)	s	p	Reference
Cd	593.9	2	2,550	2,540	12	0	c, d
Sn	504.9	4	2,510	2,540	12	0	a, b, c, d
Gd	1,585	1	2,450	2,540	12	0	f
Sb	1,653	1	2,200	2,300	12	1	d
Mg	923	4	2,050	2,070	13	0	a, b, c, d
La	1,099	1	1,700	1,690	14	0	f
Bi	544	1	1,600	1,530	14	1	d
Nd	1,113	1	1,550	1,530	14	1	f
Pr	1,213	1	1,530	1,530	14	1	f
Sm	1,573	1	1,390	1,381	15	0	f
Ce	1,077	1	1,380	1,381	15	0	f
NaCl	1,074	1	1,360	1,381	15	0	c
Ca	1,111	1	800	830	17	1	a
Sr	1,041	1	800	830	17	1	a
Yb	2,073	1	754	750	18	0	f
Pb	600.4	3	750	750	18	0	a, b, c
Tl	576	1	370	370	21	1	a
Cs	301.7	1	340	340	22	0	a

a COTTRELL, ALAN HOWARD: The Mechanical Properties of Matter. New York: John Wiley & Sons, Inc. 1964.

b Smithells Metals Reference Book, 3rd Ed., Vol. II. Butterworth & Co. (Publ.) Limited, p. 614 (1962).

c HEARMON, R. F. S.: Advances in Physics 5, 370 (1955).

d HEARMON, R. F. S.: Applied Anisotropic Elasticity. Oxford University Press 1961.

e Handbook of Chemistry and Physics, 43rd Ed., p. 2169. Cleveland, Ohio: The Chemical Rubber Publishing Co. 1961—1962.

f SMITH, J. F., C. E. CARLSON, and F. H. SPEDDING: J. Metals, Trans. 9, 1212 (1957).

g ZUCKER, C.: J. Acoust. Soc. Amer. 27, No. 2, 318 (1955).

h FRIEDEL, J.: Dislocations. Reading, Massachusetts: Addison-Wesley Publishing Co. 1964.

Appendix II (continued)   Table B

Element	T_m °K	$\mu(0)$ kg/mm² Experimental (from E & K)	$\mu(0)$ kg/mm² Predicted from Eq. (5.1)	s	p
Rh	2,239	15,830	15,750	3	0
Cr	2,163	9,669	9,480	5	1
Mn	1,533	8,784	8,560	6	0
Ga	303	790	750	18	0
Li	459	620	610	19	0
Na	371	552	550	19	1
Ba	998	575	555	19	1
In	429	548	555	19	1
K	335	234	240	23	1
Rb	311	77	73	29	1

Table C

Element	T_m °K	$\mu(0)$ kg/mm² Average Experimental	Room Temperature Averaged Experimental $\mu(300)$	$\mu(0)$ kg/mm² Predicted from Eq. (5.1)
Os	3,273	24,300	23,887	23,600
Ir	2,728	23,100	22,476	23,600
Re	3,443	21,400	21,079	21,300
Ru	2,773	19,500	18,993	19,300
Rh	2,239	15,830	15,213	15,730
W	3,683	15,700	15,512	15,730
Mo	2,893	15,400	15,030	15,730
Cr	2,163	9,669	9,273	9,480
Fe	1,809	8,700	8,222	8,580
Ni	1,726	8,600	8,084	8,580
Co	1,768	7,900	7,442	7,750
Ta	3,269	7,090	6,969	7,010
Pt	2,042	5,860	5,596	5,720
V	2,133	5,320	5,101	5,160
Pd	1,825	5,200	4,777	5,160
Ti	1,943	4,420	4,203	4,210
Nb	2,743	3,900	3,795	3,810
Zr	2,125	3,680	3,522	3,810
Hf	2,523	3,600	3,488	3,450
Er	1,798	3,180	3,020	3,110
Th	2,023	2,960	2,824	3,110
Y	1,763	2,830	2,670	2,810
Yb	2,073	754	720	750

References

ANDRADE, E. N. DA C., and D. A. ABOAV: The mechanical behaviour of single crystals of metals; in particular, copper. Proc. Roy. Soc. A **240**, 304 (1957).

—, and C. HENDERSON: The mechanical behaviour of single crystals of certain face-centred cubic metals. Phil. Trans. **244**, 177 (1951).

BARRETT, CHARLES S.: Structure of Metals. New York: McGraw-Hill 1952.

—, and L. H. LEVENSON: The structure of aluminum after compression. Transactions AIME **137**, 112 (1940).

BATEMAN, CATHERINE M.: Residual lattice strains in plastically-deformed aluminum. Acta Metallurgica **2**, No. 1, 451 (1954).

BECHTOLD, J. H., E. T. WESSEL, and L. L. FRANCE: Mechanical behavior of the refractory metals. Refractory Metals and Alloys, Vol. 11, p. 25. New York: Interscience Publishers 1961.

BELL, JAMES F.: Propagation of plastic waves in pre-stressed bars. U.S. Navy Technical Report No. 5. Baltimore: The Johns Hopkins University 1951.

— 10,000 threads to the inch. Amer. Machinist **100**, No. 16, 112 (1956a).

— Determination of dynamic plastic strain through the use of diffraction gratings. J. Appl. Physics **27**, No. 10, 1109 (1956b).

— Normal incidence in the determination of large strain through the use of diffraction gratings. Proceedings 3rd U.S. Nat'l Congress of Applied Mechanics. Brown University, Providence, Rhode Island, p. 489 (1958).

— Propagation of plastic waves in solids. J. Appl. Physics **30**, No. 2, 196 (1959) (see 1960b).

— Diffraction grating strain gauge. Proc. Soc. Exp. Stress Anal. **17**, No. 2, 51 (1960a).

— Propagation of large amplitude waves in annealed aluminum. J. Appl. Physics **31**, No. 2, 277 (1960b).

— Study of initial conditions in constant velocity impact. J. Appl. Physics **31**, No. 12, 2188 (1960c).

— Discussion. Proceedings 2d Symposium on Naval Structural Mechanics, p. 485. New York: Pergamon Press 1960d.

— An experimental study of the applicability of the strain rate independent theory for plastic wave propagation in annealed aluminum, copper, magnesium, and lead. Technical Report No. 5, U.S. Army, OOR Contract No. D-36-034-ORD-2366 (1960e).

— The initial development of an elastic strain pulse propagating in a semi-infinite bar. Technical Report No. 6. U.S. Army, Ballistics Research Laboratory. Baltimore: The John Hopkins University 1960f.

— An experimental study of the unloading phenomenon in constant velocity impact. J. Mech. Physics Solids **9**, 1 (1961a).

— Experimental study of the interrelation between the theory of dislocations in polycrystalline media and finite amplitude wave propagation in solids. J. Appl. Physics **32**, No. 10, 1982 (1961b).

BELL, JAMES F.: Further experimental study of the unloading phenomenon in constant velocity impact. J. Mech. Physics Solids 9, 261 (1961c).
— Experiments on large amplitude waves in finite elastic strain. Proceedings, I.U.T.A.M. Symposium on 2d Order Effects in Elasticity, Plasticity, and Fluid Dynamics, p. 173. Haifa 1962a. [New York: Macmillan (1964)].
— Experimental study of dynamic plasticity at elevated temperatures. Exp. Mech. 2, No. 1, 1 (1962b).
— Single temperature-dependent stress-strain law for the dynamic plastic deformation of annealed face-centered cubic metals. J. Appl. Physics 34, No. 1, 134 (1963a).
— The initiation of finite amplitude waves in annealed metals. Proceedings I.U.T.A.M. Symposium on Stress Waves in Anelastic Solids, p. 166. Brown University, Providence, R.I. 1963b. [Springer (1964)].
— A generalized large deformation behaviour for face-centred cubic solids— high purity copper. Phil. Mag. 10, No. 103, 107 (1964).
— Generalized large deformation behaviour for face-centred cubic solids: nickel, aluminium, gold, silver, and lead. Phil. Mag. 11, No. 114, 1135 (1965a).
— The dynamic plasticity of metals at high strain rates: An experimental generalization. Colloquium on Behavior of Materials under Dynamic Loading. A.S.M.E., p. 19. New York 1965b.
— The relevance of dynamic finite distortion research to high energy rate forming processes. Proceedings International Symposium on High Energy Rate Forming, p. 1. Prague: 1966a.
— An experimental diffraction grating study of the quasi-static hypothesis of the split Hopkinson bar experiment. J. Mech. Physics Solids 14, 309 (1966b).
— On the direct measurement of very large strain at high strain rates. Exp. Mech. 7, No. 1, 8 (1967a).
— Theory vs. experiment for finite amplitude stress waves. Advances in Engineering Science. New York: Gordon and Breach Science Publishers, Inc. 1967b.
— An experimental study of instability phenomena in the initiation of plastic waves in long rods. Proceedings, Symposium on the Mechanical Behavior of Materials under Dynamic Loads (San Antonio, Texas). New York: Springer 1967c.
—, and ROBERT E. GREEN, jr.: An experimental study of the double slip deformation hypothesis for face-centred cubic single crystals. Phil. Mag. 15, 469 (1967).
—, and ALBERT STEIN: The incremental loading wave in the pre-stressed plastic field. J. Mécanique 1, No. 4, 395 (1962).
—, and J. H. SUCKLING: The dynamic overstress and the hydrodynamic transition velocity in the symmetrical free-flight plastic impact of annealed aluminum. Proceedings 4th U.S. Nat'l Congress of Applied Mechanics, p. 877 (1962).
—, and W. MEADE WERNER: Applicability of the Taylor theory of the poly-crystalline aggregate to finite amplitude wave propagation in annealed copper. J. Appl. Physics 33, No. 8, 2416 (1962).
BERNER, R.: Referred to in SEEGER, A. (1957): Kristallplastizität. Handbuch der Physik (S. FLUGGE, ed.) Vol. VII/2, p. 53 (1958).
— Die Temperatur und Geschwindigkeitsabhängigkeit der Verfestigung kubisch-flächezentrierter Metalleinkristalle. Z. Naturforsch. A 15, 689 (1960).

BISHOP, J. F. W., and R. HILL: A theoretical derivation of the plastic properties of polycrystalline face-centred metals. Phil. Mag. XLII, 1298 (1951).

BLEWITT, T. H., R. R. COLTMAN, and J. K. REDMAN: Defects in Crystalline Solids, p. 369. London: The Physical Society 1955.

BOAS, W., and G. J. OGILVIE: The plastic deformation of a crystal in a polycrystalline aggregate. Acta Metallurgica 2, No. 1, 655 (1954).

—, and E. SCHMID: Über die Temperaturabhängigkeit der Kristallplastizität, III. Aluminium. Z. Physik 71, 703 (1931).

BODNER, S. R., and A. ROSEN: Discontinuous yielding of commercially-pure aluminum. J. Mech. Physics Solids 15, 63 (1967).

BOLLING, G. F., L. E. HAYS, and H. W. WIEDERSICH: The plasticity of lead single crystals and the determination of stack fault energy. Acta Metallurgica 10, 185 (1962).

BRIDGMAN, P. W.: The Physics of High Pressure. London: G. Bell & Sons 1931.

CARREKER, R. P., jr.: Tensile deformation of silver as a function of temperature, strain rate, and grain size. J. Metals 9, 112 (1957).

—, and R. W. GUARD: Tensile deformation of molybdenum as a function of temperature and strain rate. J. Metals 8 (Transactions, AIME), 178 (1956).

—, and W. R. HIBBARD, jr.: Tensile deformation of high purity copper as a function of temperature, strain rate and grain size. Acta Metallurgica 1, 654 (1953).

— — Tensile deformation of aluminum as a function of temperature, strain rate, and grain size. J. Metals 9, 1157 (1957).

CLARK, D. S.: Dynamic loading of metals. Trans. Amer. Soc. Metals 46, 34 (1953).

—, and D. S. WOOD: The tensile impact properties of some metals and alloys. Trans. Amer. Soc. Metals 42, 45 (1950).

CONN, ANDREW F.: On impact testing for dynamic properties of metals, Ph. D. dissertation. Baltimore: The Johns Hopkins University 1964.

— On the use of thin wafers to study dynamic properties of metals. J. Mech. Physics Solids 13, 311 (1965a).

— An experimental study of the impact response of 2024-T351 aluminum alloy. Martin Co. Report RR-69. Baltimore: 1965b.

COTTRELL, A. H.: Dislocations and Plastic Flow in Crystals. London: Oxford University Press 1953.

COULOMB, CHARLES A.: Recherches théoretiques et expérimentales sur la force de torsion, et sur l'élasticité des fils de métal. Histoire de l'Académie des Sciences (1784).

— Mémoires de Coulomb. Collection de Mémoires Relatifs à la Physique, publiés par la Societé Française de Physique (Paris) 1, 63 (1884).

CRISTESCU, N.: On the propagation of elastic-plastic waves in metallic rods. Bulletin de l'Académie Polonaise des Sciences, Série des Sciences Techniques XI, No. 4, 183 (1963).

CRISTESCU, N.: Some problems of the mechanics of extensible strings. I.U.T.A.M. Symposium on Stress Waves in Anelastic Solids (Brown University), p. 118. Berlin, Göttingen, Heidelberg: Springer 1964.

— On the coupling of plastic waves (differential constitutive laws). Proc. Vibration Problems (Warsaw) 2, No. 6, 145 (1965a).

— Loading/unloading criteria for rate sensitive materials. Arch. Mech. Stosowanej (Bucureşti) 2, No. 17, 291 (1965b).

DAVIES, H. D. H., and S. C. HUNTER: The dynamic compression testing of solids by the method of the split Hopkinson pressure bar. J. Mech. Physics Solids 11, 155 (1963).

DIEHL, J.: Zugverformung von Kupfer-Einkristallen, I. Verfestigungskurven und Oberflächenerscheinungen. Z. Metallkunde 47, 331 (1956).

DILLON, O. W., jr.: Experimental data on aluminum as a mechanically unstable solid. J. Mech. Physics Solids 11, 289 (1963).

— Waves in bars of mechanically unstable materials. J. Appl. Mechanics 33, 267 (1966).

DOHI, SHOSO: Tensile straining and deformation bands of single crystal of pure iron. J. Science, Hiroshima University, Ser. A, 24, No. 1 (1960).

ENRIETTO, J. F., G. M. SINCLAIR, and C. A. WERT: Mechanical behavior of columbium containing oxygen. Metallurgical Society Conferences Vol. 10, p. 503 (1961).

ERICKSEN, J. L.: Oriented solids. Proceedings, Symposium on Structural Dynamics under High Impulse Loading, p. 37 (1963).

FILBEY, GORDON, L., jr.: Intense plastic waves. Ph.D. dissertation. Baltimore: The Johns Hopkins University 1961a.

— Deformation waves in annealed aluminum rods undergoing high velocity impact. Technical Report No. 8, U.S. Army Ballistics Research Laboratories, Contract No. DA-36-034-21X4992, 509-ORD-3104RD. Baltimore: The Johns Hopkins University 1961b.

— Uniaxial stress conditions. Proceedings, Symposium on Structural Dynamics under High Impulsive Loading, p. 147 (1963).

— Longitudinal plastic deformation waves in bars. Proceedings, Princeton University (U.S.A.) Conference on Solid Mechanics, p. 111 (1965).

FITZGERALD, EDWIN R.: Particle Waves and Deformation in Crystalline Solids. New York: Interscience Publishers 1966.

FUCHS, K.: A quantum mechanical calculation of the elastic constants of monovalent metals. Proc. Roy. Soc. London. Ser. A. 153, 622 (1936).

GILLICH, WILLIAM J.: The response of bonded wire resistance strain gauges to large amplitude waves in annealed aluminum. M. S. essay. Baltimore: The Johns Hopkins University 1960.

— Plastic wave propagation in high purity single crystals of aluminum. Ph. D. dissertation. Baltimore: The Johns Hopkins University 1964.

— Propagation of finite amplitude waves in single crystals of high-purity aluminum. Phil. Mag. 15, No. 136, 659 (1967).

GILMAN, J. J.: Dislocation mobility in crystals. J. Appl. Physics 36, No. 10, 3195 (1965a).

— Microdynamics of plastic flow at constant stress. J. Appl. Physics 36, No. 9, 2772 (1965b).

GOLER, F. V., and G. SACHS: Das Verhalten von Aluminiumkristallen bei Zugversuchen. I. Geometrische Grundlagen. Z. Physik 41, 103 (1927).

GREENOUGH, G. B.: Quantitative x-ray diffraction observations on strained metal aggregates. Progr. Metal Physics 3, 176 (1952).

HAASEN, P.: Plastic deformation of nickel single crystals at low temperatures. Phil. Mag. 3, 384 (1958).

HARTMAN, WILLIAM F.: The applicability of the generalized parabolic deformation law to a binary alloy. Ph.D. dissertation. Baltimore: The Johns Hopkins University 1967.

HEINRICHS, J. A.: An experimental study of 0.483 inch diameter aluminum bars under constant velocity impact. M.S. essay. Baltimore: The Johns Hopkins University 1961.

HOCKETT, J. E.: Compression testing at constant true strain rates. Proc. Amer. Soc. Testing Materials 59, 1309 (1959).

HOPKINSON, B.: A method of measuring the pressure produced in the detonation of explosives or by the impact of bullets. Phil. Trans. Roy. Soc. A 213, 375 (1914).

HOPKINSON, JOHN: On the rupture of iron wire by a blow. In: Original Papers by the Late John Hopkinson, Vol. II, p. 316. B. HOPKINSON, ed. Cambridge: Cambridge University Press 1901.

HOSFORD, W. F., R. L. FLEISCHER, and W. A. BACKOFEN: Tensile deformation of aluminum single crystals at low temperatures. Acta Metallurgica 8, 187 (1960).

JAFFEE, R. I., D. J. MAYKUTH, and R. W. DOUGLAS: Rhenium and the refractory platinum-group metals. Refractory Metals and Alloys, Metallurgical Society Conferences Vol. 11, p. 383 (1961).

JOHNSON, J. E., D. S. WOOD, and D. S. CLARK: Dynamic stress-strain relations for annealed 2S aluminum under compression impact. J. Appl. Mechanics 20, 523 (1953).

KARMAN, THEODORE, VON: On the propagation of plastic deformation in solids. NDRC Report A29, OSRD 365 (1942).

KARNOP, R., and G. SACHS: Das Verhalten von Aluminiumkristallen bei Zugversuchen. Z. Physik 41, 116 (1927).

KELLY, A.: The mechanism of work softening in aluminum. Phil. Mag. 8th ser. 1, 835 (1956).

KELVIN, LORD WILLIAM THOMSON: Elasticity. Encyclopedia Britannica, 9th ed., 9, 796 (1880).

KENIG, M. J., and O. W. DILLON, jr.: Shock waves produced by small stress increments in annealed aluminum. J. Appl. Mechanics 33, 907 (1966).

KINGMAN, P. W., R. E. GREEN, and R. B. POND: Reaction of fine metal wires to imposed loads. U.S. Army Terminal Report, Ballistics Research Laboratories, Aberdeen Proving Ground, Maryland (1963).

KRONMÜLLER, H.: Das magnetische Einmündungsgesetz bei plastisch verformten Nickel- und Nickel-Kobalt-Einkristallen. Z. Physik 154, 574 (1959).

KRUPNIK, N., and HUGH FORD: The stepped stress/strain curve of some aluminum alloys. J. Inst. Metals 8, 601 (1953).

KUHLMANN-WILSDORF, D.: Unified theory of stages II and III of workhardening in pure FCC metal crystals. Symposium on Workhardening, Chicago, Illinois. New York: Gordon and Breach 1967.

LUBLINER, JACOB: A generalized theory of strain-rate dependent plastic wave propagation in bars. J. Mech. Physics Solids 12, 59 (1964).

LÜCKE, K., and H. LANGE: Über die Form der Verfestigungskurve von Reinstaluminiumkristallen und die Bildung von Deformationsbändern. Z. Metallkunde 43, 55 (1952).

MADER, S., A. SEEGER, and C. LEITZ: Work hardening and dislocation arrangement of FCC single crystals. I. Plastic deformation and slip line studies of nickel single crystals. J. Appl. Physics 34, 3368 (1963).

MALVERN, L. E.: The propagation of longitudinal waves of plastic deformation in a bar of material exhibiting a strain rate effect. J. Appl. Mechanics 18, 203 (1951).

MANJOINE, M. J., and A. NADAI: High speed tension tests at elevated temperatures. Proc. Amer. Soc. Testing Materials **40**, 822 (1940).

MARCINKOWSKI, M. J., and H. A. LIPSITT: The plastic deformation of chromium at low temperatures. Acta Metallurgica **10**, 95 (1962)

McKEOWN, J., and O. F. HUDSON: Characteristics of copper, silver, and gold. J. Inst. Metals LX, No. 1, 109 (1937).

McREYNOLDS, A. W.: Plastic deformation waves in aluminum. Trans. Amer. Inst. Mining Metallurgical Engineers **185**, 232 (1949).

MEISSNER, J., VON: Untersuchungen über die Plastizität von Einkristallen der Legierungsreihe Nickel-Kobalt. Z. Metallkunde **50**, 207 (1959).

MITCHELL, T. E., and W. A. SPITZIG: Three-stage hardening in tantalum single crystals. Acta Metallurgica **13**, Part 2, 1169 (1965).

—, and P. R. THORNTON: The working-hardening characteristics of Cu and α-brass single crystals between 4.2 and 500° K. Phil. Mag. **8**, 1127 (1963).

MOTT, N. F.: A theory of work-hardening of metal crystals. Phil. Mag. 7th series **43**, 1151 (1952).

—, and H. JONES: The Theory of the Properties of Metals and Alloys. New York: Oxford Press 1936.

MURPHY, HEATHER M., and E. A. CALNAN: The deformation of single crystals of α-brass. Acta Metallurgica **3**, 268 (1955).

NOGGLE, T. S., and J. S. KOEHLER: Electron microscopy of aluminum crystals deformed at various temperatures. J. Appl. Physics **28**, 53 (1957).

PARKER, EARL R., and THOMAS H. HAZLETT: Principles of solution hardening. Seminar on Relation of Properties to Microstructure, American Society of Metals (1954).

PATEL, J. R., and B. H. ALEXANDER: Plastic deformation of germanium in compression. Acta Metallurgica **4**, 385 (1956).

PHILLIPS, A. J., and A. A. SMITH: Effect of time on tensile properties of hard-drawn copper wire. Proc. ASTM **36**, Pt. 2, 263 (1936).

PHILLIPS, V. A.: Effect of composition and heat-treatment on yield-point phenomena in aluminum alloys. J. Inst. Metals **8**, 649 (1953).

— A. J. SWAIN, and R. EBORALL: Yield-point phenomena and Stretcher-strain markings in aluminum-magnesium alloys. J. Inst. Metals **81**, 625 (1953).

POND, R. B., and ELEANOR HARRISON: Grain boundary movement in bicrystalline aluminum. Trans. Amer. Soc. Metals **50**, 994 (1958).

PORTEVIN, A., and M. A. LE CHATELIER: Sur un Phénomène Observé lors de l'Essai de Traction d'Alliages en cours de Transformation. Comptes Rendus **176**, 507 (1923).

PRICE, R. J., and A. KELLY: Deformation of age-hardened crystals of copper—1.8% beryllium. Acta Metallurgica **11**, 915 (1963).

— — Deformation of age-hardened aluminum alloy crystals—I. Plastic flow. Acta Metallurgica **12**, 159 (1964).

RAKHMATULIN, K. A.: Propagation of a wave of unloading. Sov. J. Appl. Mathematics Mechanics (Prikl. Mat. Mekh.) **9**, No. 1, 91 (1945).

RICHARDS, JOHN T.: An evaluation of several static and dynamic methods for determining elastic moduli. Symposium on Determination of Elastic Constants, 55th Annual Meeting, American Society for Testing Materials, Technical Publication No. 129, p. 71. New York 1952.

RICHARDS, T. W.: Concerning the compressibility of the elements and their relations to other properties. J. Amer. Chem. Soc. **37**, Pt. 2, 1643 (1915).

ROSEN, A., and S. R. BODNER: The influence of strain rate and strain ageing on the flow stress of commercially pure aluminum. Scientific Report No. 1, Contract AF61(052)-951, Israel Institute of Technology (1966).

ROSENTHAL, D., and W. B. GRUPEN: Second order effect in crystal plasticity: deformation of surface layers in face-centred cubic aggregates. I.U.T.A.M. Internat'l Symposium on Second-Order Effects in Elasticity, Plasticity, & Fluid Dynamics. Haifa 1962. [New York: Macmillan (1964).]

SACHS, G., and J. WEERTS: Zugversuche an Gold-Silberkristallen. Z. Physik **62**, 473 (1930).

SEEGER, A., J. DIEHL, S. MADER, and H. REBSTOCK: Work-hardening and work-softening of face-centred cubic metal crystals. Phil. Mag. **2**, 323 (1957).

SHARPE, W. N., jr.: The Portevin - le Chatelier effect in aluminum single crystals and polycrystals. Ph.D. dissertation. Baltimore: The Johns Hopkins University 1966a.

— An experimental investigation of the Portevin - le Chatelier effect in dead annealed commercial purity aluminum. U.S. Air Force Report, AFOSR 66-0127, Contract No. AF 49(638)-1067. Baltimore: The Johns Hopkins University 1966b.

— The Portevin - le Chatelier effect in aluminum single crystals and polycrystals. J. Mech. Physics Solids **14**, 187 (1966c).

SIMMONS, C. R.: The mechanical properties of yttrium, scandium, and the rare-earth metals. The Rare Earths. New York: Wiley & Sons 1961.

SMITH, J. F., C. E. CARLSON, and F. H. SPEDDING: Elastic properties of yttrium and eleven of the rare-earth elements. Trans. J. Metals **9**, Pt. 10, 1212 (1957).

SOKOLOVSKY, V. V.: The propagation of elastic-visco-plastic waves in bars. Sov. J. Appl. Mathematics Mechanics (Prikl. Mat. Mekh.) XII, No. 3, 261 (1948).

SPERRAZZA, J.: Propagation of large amplitude waves in pure lead. Dr. Eng. dissertation. Baltimore: The Johns Hopkins University 1961.

— Propagation of large amplitude waves in pure lead. Proceedings, 4th U.S. Nat'l Congress Applied Mechanics Vol. 2, p. 1123 (1962a).

— Propagation of large amplitude waves in pure lead. U.S. Army Ballistics Research Laboratories. Aberdeen Proving Ground, Maryland. Report No. 1158, Project No. 503-04-002 (1962b).

STAUBWASSER, W.: Doctoral Dissertation, Göttingen. [Reference in SEEGER, A. (1954), Kristallplastizität, Handbuch der Physik, edited by S. FLÜGGE, Vol. VII/2 (1958).]

STEIN, ALBERT: An experimental study of incremental plastic wave propagation. M. S. essay. Baltimore: The Johns Hopkins University 1962.

SUZUKI, H., S. IKEDA, and S. TAKEUCHI: Deformation of thin copper crystals. J. Phys. Soc., Japan **11**, 382 (1956).

TAYLOR, G. I.: The mechanism of plastic deformation of crystals. Proc. Roy. Soc., Ser. A **145**, 362, 388 (1934). [Also in Scientific Papers of Sir Geoffrey Ingram Taylor, edited by G. K. BATCHELOR, p. 344. (Cambridge: Cambridge University Press 1958).]

— Plastic strain in metals. J. Inst. Metals LXII, 307 (1938). [In Scientific Papers of ... (*see supra*), p. 424.]

TAYLOR, G. I.: The plastic wave in a wire extended by an impact load. British Official Report, RC 329 (1942).

—, and C. F. ELAM: The plastic extension and fracture of aluminum crystals. Proc. Roy. Soc. A 108, 28 (1925).

—, and H. QUINNEY: The distortion of wires on passing through a drawplate. J. Inst. Metals XLIX, 187 (1932).

THIELE, WALTER: Temperaturabhängigkeit der Plastizität und Zugfestigkeit von Steinsalzkristallen. Z. Physik. LXXV, 763 (1932).

THORNTON, P. R., T. E. MITCHELL, and P. B. HIRSCH: The strain-rate dependence of the flow stress of copper single crystals. Phil. Mag. 7, 337 (1962).

TRUESDELL, C.: General and exact theory of waves in finite elastic strain. Arch. Rational Mechanics Analysis 8, No. 3, 263 (1961).

WERTHEIM, G.: Recherches sur l'Élasticité; Premier Mémoire. Ann. Chimie (Paris) XII, 385 (1844a).

— Recherches sur l'Élasticité: Dieuxième Mémoire. Ann. Chimie (Paris) XII, 581 (1844b).

— Untersuchungen über die Elasticität. Ann. Physik u. Chemie (Poggendorff), Bd. II, I: p. 1; II: p. 73. Leipzig: 1848.

WHITE, M. P., and LeVAN GRIFFIS: The permanent strain in a uniform bar due to longitudinal impact. N.D.R.C. (Progress Report No. A-7), O.S.R.D. No. 742 (1942).

— — The propagation of plasticity in uniaxial compression. J. Appl. Mechanics 15, 256 (1948) (Trans., ASME 70, No. 256).

YOUNG, THOMAS: On the equilibrium and strength of elastic substances. Section No. 46 from Miscellaneous Works of Thomas Young, p. 129. London: S. Murray 1855. Reprinted from Mathematical Elements of Natural Philosophy, Vol. II of Dr. Young's Lectures, Section IX, p. 46 (1807).

ZUCKER, C.: Elastic constants of aluminum from 20° C to 400° C. J. Acoust. Soc. Amer. 27, No. 2, 318 (1955).

Author Index

Page numbers in *italics* refer to the references

Subject Index

Springer Tracts in Natural Philosophy